EUREKA!

SCIENCE'S GREATEST THINKERS AND THEIR KEY BREAKTHROUGHS

EUREKA!

SCIENCE'S GREATEST THINKERS AND THEIR KEY BREAKTHROUGHS

HAZEL MUIR

Quercus

INTRODUCTION

Eureka! charts the lives and work of 300 scientists who have made a significant, often dramatic, impact on the world. Some of them have benefitted society in direct and obvious ways, like developing live-saving drugs or world-changing technologies. Others have had remarkable insights about the basic laws of nature or the Earth's restless past.

From the 1600s, modern scientific thinking emerged with an emphasis on thorough experimentation and evidence. Scientists were spurred on by the brand new tools of the trade – the telescope, revealing vast new worlds and galaxies, and the compound microscope, making the invisible visible.

The concise biographies in this book show what this change looked like, with woolly ancient speculations, often framed by religious dogma or mysticism, giving way to more rational argument. Each entry is organized chronologically according to when the scientist produced their most significant work; each concentrates on a specific contribution and its impact, rather than an exhaustive list of prizes and honorary degrees.

My selection focuses mainly on natural science (biology, physics and chemistry and their related fields), as well as mathematics, social science and archaeology. Selecting living scientists proved especially tough. Many great achievements of the modern era have been made by large teams of people, and it's neither possible, nor fair, to credit one or two individuals. There are also many scientists currently doing exciting work on complex problems, but it's not clear, even to them, where their work will lead. This makes it difficult to assess their future impact.

Even in science, fame doesn't rest solely on academic or technical achievement. Scientists can be renowned for their vigorous campaigning on ethical issues, their communication skills or their eccentricity, as well as their scientific work. And they're not immune to disrepute – Trofim Lysenko's (p. 253) chief claim to fame, for example, was fraudulent work on agriculture and a ruthless persecution of his peers.

Other people have become famous for things they didn't set out to do. Who would have heard of Charles Messier (p. 66), had his name not been linked to the most stunning objects in the night sky? Messier unwittingly catalogued beautiful exploded stars and magnificent galaxies – but to him they were just annoying fuzzy patches that got in the way of his comet hunting.

The lasting fame of many other scientists rests partly on having their name immortalized, for example in some unit of measurement. Many of these are included too, as well as some who are simply interesting and unusual.

Eureka! doesn't claim to represent a definitive selection of the people who have made the most important contributions to science, and my apologies if any of the big names have slipped the net. But hopefully, the selection makes for a good read, and conveys the awesome ingenuity of which humanity is capable.

Hazel Muir, 2012

PYTHAGORAS
c.570 – c.495 BC

Ancient Greek philosopher and mathematician who founded the Pythagorean religious sect

'Number is the ruler of forms and ideas, and the cause of gods and daemons'

Pythagoras quoted in *Life of Pythagoras* (c.300 AD)
by Assyrian philosopher Iamblichus of Chalcis

Born on the island of Samos, Pythagoras was a Greek philosopher and mathematician best known for the theorem that bears his name. As with many philosophers of his era, little is known about his life except for fragmentary information passed on by later philosophers, who say that Pythagoras moved to the Greek colony of Croton (now Crotone in southern Italy) around 530 BC. He founded a religious cult there, but eventually fled the city when the sect attracted hostility from outsiders. Many members of the cult died when their Pythagorean meeting place was torched.

The 3rd-century Greek anthologist Diogenes Laërtius suggested Pythagoras travelled in search of knowledge as far and wide as Egypt, Babylon and India. He discovered Pythagoras's theorem, which states that the square of the length of the hypotenuse (longest side) of a right-angled triangle is equal to the sum of the squares of lengths of the two other sides. The Babylonians knew the theorem as much as 1,000 years earlier, but Pythagoras may have been the first person to prove it. He or one of his followers is said to have discovered irrational numbers – numbers such as the square root of 2 that can't be expressed as a fraction of two whole numbers (such as 5/2).

The Pythagorean sect was profoundly secretive and members treated certain symbols with mystical significance, swearing oaths by the 'tetractys' (a triangular shape with four rows of dots, adding up to the 'perfect number' ten). Some accounts suggest the cult promoted vegetarianism and forbade members from eating beans or drinking wine, but it's largely impossible to distinguish legend from reality.

ANAXIMANDER
c.610 – c.546 BC

Ancient Greek early proponent of science who questioned the origins of life and the Universe and believed that nature is ruled by laws

'Anaximander of Miletus conceived that there arose from heated water and earth either fish or creatures very like fish'

How the 3rd-century Roman writer Censorinus described Anaximander's view of evolution

Anaximander was born in Miletus, an ancient Greek city on the west coast of Anatolia (now western Turkey). Very little is known about his life. He wrote several works but only a fragment of one, *On Nature*, survives. What we know of him comes largely from snippets of information passed on by later philosophers.

On Nature was a philosophical poem in which Anaximander speculates on the origins of the Earth as well as plants and animals. He was aware of the existence of fossils and might be viewed as the earliest proponent of evolution. He believed life evolved from fish-like creatures that inhabited a watery world in ancient days before the Sun dried the land.

Anaximander suggested the first humans evolved inside fish-like animals 'held prisoners until puberty', according to the 3rd-century Roman writer Censorinus. 'Only then, after these animals burst open, could men and women come out, now able to feed themselves.' Although these speculations sound bizarre, they nonetheless represent a scientific attempt to explain human origins in a way that was rational rather than resorting to mysticism or religion.

He is also credited with developing the first mechanical model of the world. He speculated that the Earth's surface is curved, although he envisioned its shape as a cylinder rather than a sphere, the flat top possibly representing the inhabited world. The stars circled around the Earth, with the Moon farther away and the Sun most distant of all. He believed there might be an infinite number of inhabited worlds, all undergoing cycles of dissolution and renewal.

DEMOCRITUS
*c.*460 – *c.*370 BC

Ancient Greek philosopher famous for his
atomic theory of the cosmos

'Nothing exists except atoms and empty space;
everything else is opinion'

Democritus quoted in Diogenes Laërtius's *Lives and Opinions of Eminent Philosophers,*
3rd century AD

Democritus was born into a wealthy family in the city of Abdera, Thrace. He travelled widely in search of knowledge, possibly visiting Egypt, Ethiopia, Persia and India. Diogenes Laërtius, a 3rd-century biographer of the Greek philosophers, reported that he had a cheerful and modest personality, and devoted all his time to studying.

Democritus was said to be a prolific writer on ethics, politics, maths, biology and cosmology, but only fragments of his works survive; most of his ideas were passed on by later philosophers. They say he had a rationalist philosophy and believed everything is the result of natural laws. He held that matter is composed of indivisible 'atoms' with empty space in between. Although that sounds like a modern scientific concept, there was no solid experimental evidence for this during Democritus's lifetime – that only emerged more than 2,000 years later.

The world consisted of an infinite number of indestructible particles, according to Democritus. Everything is already in the world ('nothing is created out of nothing') – change results only from combination and recombination of eternal particles that can combine using various connections like hooks and barbs.

Democritus perceived the Earth was round and suggested that the Universe began as a churning mass of tiny atoms that gradually clumped together to form worlds like the Earth. He was open to the idea that there are many worlds that grow and decay in an endless repeating cycle. In maths, Democritus was the first to determine that the volume of a cone is one-third the volume of a cylinder with the same base size and height.

HIPPOCRATES
c.460 – c.370 BC

*Greek physician who is considered the
father of rational medicine*

'As to diseases, make a habit of two things – to help,
or at least to do no harm'

From Hippocrates's *Epidemics* (vol 1)

 **Born on the Greek island of Kos**,
Hippocrates was probably taught by his father and
grandfather, both physicians. He later founded a
medical school on the island. He travelled widely
to examine patients and was very famous in his
own lifetime, but little else is known about him.
Some accounts suggest that he clashed with Greek
authorities over his theories and approaches, and
was sentenced to two decades in prison as a result.

He may have been a prolific writer – the *Corpus Hippocraticum* (the
Hippocratic Collection) numbers around 70 works, although as they
were written over a period longer than his lifespan, clearly other people
contributed to them. His *Epidemics* gave clear and rational descriptions of
the causes, development and prognosis of illnesses with any discussion of
therapies tending to focus on diet, cleanliness and exercise.

Hippocrates is thought to be the first influential person who
promoted the idea that disease does not arise from mystical causes, such
as being inflicted by gods – instead, he recognized that it arises naturally,
with diet, lifestyle and environmental factors such as the weather playing
a role. He also made extremely detailed observations of diseases of the
respiratory tract.

However, Hippocrates did subscribe to some erroneous contemporary
notions, such as 'humourism'. It was popular to believe that the human
body is filled with four humours (black bile, yellow bile, phlegm and blood)
and that good health requires them to have the right relative amounts. He
or one of his contemporaries wrote an early version of the 'Hippocratic
oath', a vow often taken by doctors to practise medicine ethically, to treat
patients with respect and honour their privacy.

ARISTOTLE
384 – 322 BC

Greek philosopher whose cosmological ideas dominated for 18 centuries

'Those who assert that the mathematical sciences say nothing of the beautiful or the good are in error ... The chief forms of beauty are order and symmetry and definiteness, which the mathematical sciences demonstrate in a special degree.'

From Aristotle's *Metaphysics*, book XIII

Born in Stageira, northern Greece, Aristotle was a major figure in Greek philosophy. His father was court physician to the Macedonian royal family, and he trained in medicine before moving to Athens to study philosophy with Plato. He later spent time travelling and possibly studying biology in Asia Minor (now Turkey).

Around 343 BC, Aristotle was invited by Philip II of Macedon to become tutor to his son, Alexander the Great. He was also head of the royal academy of Macedon and tutored two other future Macedonian kings, Ptolemy I Soter and Cassander. After Alexander became king of Macedonia in 336 BC, Aristotle returned to Athens to found his own school, the Lyceum. When Alexander died, however, anti-Macedonian sentiment flared in Athens and Aristotle was accused of impiety. He fled to the island of Euboea where he died soon after.

Aristotle wrote around 150 philosophical treatises, some of which may have been sets of lecture notes. About 30 survive, and they span a huge range of philosophical topics, from biology and physics to ethics and politics. He was an excellent naturalist. His work in zoology, although flawed, constituted the most ambitious biological survey of its time. He wrote in detail about around 500 different animals, including 120 types of fish and 60 kinds of insect.

Aristotle distinguished whales and dolphins from fish and studied the social organization of bees, as well as the embryological development of chickens and the chambered stomachs of ruminants. He described the anatomy of many marine invertebrates including octopuses and cuttlefish.

The detail suggests he personally carried out dissection on his specimens and his texts remained influential for many centuries after his death.

In a treatise called *Meteorology*, Aristotle discussed a vast range of natural phenomena, from weather patterns to the nature of the Earth, rainbows, thunder and lightning, comets, shooting stars and the Milky Way. He challenged earlier theories that suggested the repeated wetting and drying of land triggers earthquakes, which he instead put down to strong winds inside the Earth.

His *Meteorology* reveals that he grasped our planet's basic water cycle: 'Now the Earth remains but the moisture surrounding it is made to evaporate by the Sun's rays and the other heat from above, and rises. But when the heat which was raising it leaves it, in part dispersing to the higher region, in part quenched through rising so far into the upper air, then the vapour cools because its heat is gone and because the place is cold, and condenses again and turns from air into water. And after the water has formed it falls down again to the Earth.'

Aristotle developed a theory of motion, defining concepts of 'force' and 'natural motion'. His main astronomical work was *On the Heavens*. His world view was that the Earth is the centre of the Universe and motions on it are linear and finite, while the heavenly bodies endlessly travel in perfect circles. His views on matter were somewhat backward compared to those of the atomists such as **Democritus**. A body's motion was a consequence of its composition; the stars and planets were made of a perfect element Aristotle called 'ether', while objects on the Earth contained various mixtures of the four ancient elements, earth, air, fire and water.

Despite his erroneous beliefs, Aristotle is considered one of the fathers of modern science because in all his works he emphasized objectivity, reason and rational argument. However, ironically, his legacy was to inhibit the pace of science for many centuries to come. His many works were translated into Arabic and Latin, and even some of his deeply flawed assumptions went largely unquestioned for nearly 2,000 years. His views of the physical Universe served as the key guide for natural philosophers until the 16th and 17th centuries, when **Copernicus**, **Galileo**, **Newton** and others revolutionized science.

EUCLID OF ALEXANDRIA

C.325 BC – UNKNOWN

One of the most prominent mathematicians in antiquity,
often called the father of geometry

'Someone who had begun to read geometry with Euclid, when he
had learnt the first theorem, asked Euclid, "But what shall I get by
learning these things?" Euclid called his slave and said "Give him
threepence, for he must make gain out of what he learns."'

Tale recounted by 5th-century Greek anthologist Joannes Stobaeus

Little is known about Euclid, except that
he taught in Alexandria, where he appears to have
founded a school of mathematics. It's likely that
among other works, he wrote the entire 13-volume
work the *Elements*. However, it's possible that other
mathematicians contributed to the treatise, or even
that Euclid didn't exist and was simply a pen-name
other mathematicians used when they wrote the work.
Whatever the answer, the *Elements* is the earliest substantial
mathematical treatise to survive and altered editions of it were still used
as school textbooks in the early 20th century. Subjects covered include
Pythagoras's theorem as well as the geometry of lines in a plane, stating that
only one parallel line can be drawn through a given point parallel to another
straight line. That statement underpins 'Euclidean geometry', the only type
of geometry considered until the early 19th century when mathematicians
started to explore the more complex geometry of curved spaces.

The *Elements* also discusses older number theory, including the proof
that there must be an infinitely large number of prime numbers, numbers
that have no whole-number factors except themselves and one. Euclid also
wrote surviving books on optics and astronomy. The sarcastic quote at the
top of this page, if genuine, suggests he was passionate about the value of
knowledge for its own sake.

ARISTARCHOS OF SAMOS

c.310 – c.230 BC

Greek mathematician and astronomer who proposed a Sun-centred Universe and made pioneering attempts to determine the distances of the Sun and Moon

'His hypotheses are that the fixed stars and the Sun remain unmoved, that the Earth revolves about the Sun on the circumference of a circle, the Sun lying in the middle of the floor'

Archimedes describes the heliocentric system of Aristarchos in his brief work *The Sand Reckoner*

Aristarchos was born on the Greek island of Samos. He developed the earliest known heliocentric cosmology that placed the Sun, rather than the Earth, at the centre of the Universe. But this idea did not gain acceptance until around 18 centuries after his death. Only one of Aristarchos's works survives, *On the Sizes and Distances of the Sun and Moon*. This describes his experiments to estimate the distances to the Sun and Moon by assuming (correctly) that a half Moon appears half lit because it approximately forms a right angle with the Sun and the Earth.

He concluded from his measurements that the Sun is about 18 to 20 times more distant than the Moon, which was wildly inaccurate (the real answer is about 390 times farther), but nonetheless this represented the first reasoned attempt to measure astronomical distances. He also noted that the Sun and Moon have the same angular size in the sky. So if the Sun were 18 to 20 times more distant than the Moon, the Sun must also be 18 to 20 times wider than the Moon. The argument was correct, even though the numbers were wrong.

In his short text *The Sand Reckoner*, **Archimedes** records that Aristarchos suggested the Earth moves around the Sun and the stars are very far away. He also put the planets known at the time in the correct order of distance from the Sun.

ARCHIMEDES
287 – 212 BC

Greek scientist, mathematician and engineer who formulated Archimedes's Principle

'Eureka!'

Legend says Archimedes shouted this as he ran down the street naked

 **Archimedes was surely the greatest** mathematician and scientist of ancient times, inspired to learn science by his father, an astronomer. He was born in Syracuse (now Sicily), where he studied in Alexandria under the mathematician Conon, who had earlier been a pupil of **Euclid**.

Archimedes is most famous for formulating Archimedes's Principle, which states that the buoyant force on an object submerged in a fluid (liquid or gas) is equal to the weight of the fluid that the object has displaced. It implies that an object will sink in a fluid if its average density is greater than that of the fluid. The principle explains why ships float and why hot air balloons rise, the warm air inside a balloon being less dense than the cooler air outside.

Archimedes was a very hands-on experimenter in mechanics who loved to put his discoveries to practical use. He devised a theory of levers, and, amazed by their power to move heavy weights, he remarked, 'Give me a place to stand on, and I will move the Earth', according to the mathematician Pappus of Alexandria. One story suggests that Archimedes rigged up a set of compound levers to impress Hieron II, king of Syracuse from 270 to 215 BC. He then used it to pull a fully laden ship out of a harbour and onto dry land, if the story is to be believed.

Archimedes's many inventions included the Archimedes screw, a screw-shaped blade inside a cylinder that could be hand-turned to remove water from ships, or raise water from low-lying reserves into irrigation channels. In mathematics, he calculated the value of π (pi) with unprecedented accuracy. To do this, he calculated the sizes of polygons with ever increasing numbers of sides, recognizing that with more sides a polygon will approach a circle in shape and area. In concept, this anticipated the methods of calculus developed by **Newton** and **Leibniz** nearly 2,000 years later, and some experts believe Archimedes would have developed calculus if the decimal number system had been in use at the time.

In his short manuscript *The Sand Reckoner*, Archimedes attempted to calculate the number of grains of sand that the Universe could contain, making some guesses about how big the Universe was. His main purpose was to challenge the notion that the number of grains of sand could ever be too large to be counted – in other words, a finite Universe cannot contain an infinite, immeasurable amount of anything.

But Archimedes is most famous for an anecdote recounted by later historians about how he formulated the principle named after him. Their accounts suggest that Hieron II tasked him with determining whether a crown crafted supposedly from pure gold also contained some cheaper silver. He was strictly warned not to damage the precious crown in the process.

While taking a bath, Archimedes noticed that the water level rose when he got in. It made him realize that by placing the crown in water and measuring the displaced water volume, he could establish the volume of the crown and then calculate its density and purity compared to pieces of pure gold. Legend has it that Archimedes then excitedly ran down the street naked shouting 'Eureka!', Greek for 'I have found it!'. Supposedly, the crown turned out to be partly silver and the goldsmith responsible was executed.

During the Siege of Syracuse in 214–212 BC, weapons developed by Archimedes, including powerful catapults, were used to protect the city against the Romans and at least delayed their attack. One weapon he is said to have constructed is the 'claw of Archimedes' or 'ship shaker', a big metal grappling hook hanging from a crane that could lift attacking ships out of the water then sink them. But Archimedes himself was killed by a Roman soldier, despite orders from the Roman general Marcellus to spare his life.

HIPPARCHOS
(SOMETIMES HIPPARCHUS)
C.190 – C.120 BC

Greek astronomer who founded trigonometry and turned Greek astronomy into a practical, predictive science

'It is a vulgar belief that our astronomical knowledge dates only from the recent century when it was rescued from the monks who imprisoned Galileo; but Hipparchus ... who among other achievements discovered the precession of the eqinoxes, ranks with the Newtons and the Keplers'

Written by Benjamin Disraeli in *Lothair* (1879)

Born in Nicaea (now İznik in Turkey), Hipparchos seems to have made astronomical observations from the ancient province of Bithynia (corresponding roughly to north and central Turkey), Rhodes and Alexandria. Little of this work has survived; most of what we know about him comes from later writers, particularly **Ptolemy**, who frequently refers to him with admiration in his *Almagest*.

Hipparchos created the first known star catalogue by measuring the positions of around 850 stars with unprecedented accuracy. Scientists were still making use of his catalogue as late as the 1600s. He also developed an early form of trigonometry by tabulating the length of a 'chord' joining two points on a circle depending on the angle the chord subtends at the circle's centre. Effectively, he compiled a sine table.

Probably his best-known discovery is the precession of the equinoxes. His accurate measurements of the positions of the stars alerted him to the fact that the Earth's axis of rotation gradually shifts like a wobbling top, tracing out a complete circle over time. This cycle repeats every 26,000 years. Hipparchos is also thought to have been the first person to devise a Sun-centred model of the solar system, although he discarded it because it was mathematically incompatible with circular orbits, assumed to be essential at the time. He calculated the length of the year to within 6.5 minutes and also introduced the division of a circle into 360 degrees into Greece.

LUCRETIUS
(TITUS LUCRETIUS CARUS)
c.95 BC – c.55 BC

Roman philosopher and poet who opposed religion and advocated scientific thinking

'Then be it ours with steady mind to clasp
The purport of the skies – the law behind
The wandering courses of the Sun and Moon;
To scan the powers that speed all life below;
But most to see with reasonable eyes
Of what the mind, of what the soul is made.'

From Lucretius's *On the Nature of Things* (*c.*56 BC)

Little is known about Lucretius's life. His only known work is an epic poem *De Rerum Natura* (translated into English as *On the Nature of Things*) intended to relay to a Roman audience the ideas of the Greek philosopher Epicurus who lived from 341 to 270 BC. Epicurus was a prolific writer but practically none of his works survived; Lucretius was an ardent follower whose epic poem survived intact. One copy is kept at the Laurentian Library in Florence, Italy.

In line with **Democritus**, Lucretius believed that all matter was composed of atoms. In his view, even immaterial things like the soul were made up of atoms, although these atoms were finer than those in material objects. Lucretius argued that the Universe is infinite and contains an infinite number of atoms, and that in all probability, there must be other worlds like the Earth, and such worlds must evolve then disappear. Both these concepts ran counter to religious views of the Earth being the product of divine creation.

Lucretius did not believe in the concept of the afterlife, and like Epicurus, he hoped his work would relieve people from the burden of religious fear.

CLAUDIUS PTOLEMY
c.90 – c.168 AD

Greek or Egyptian mathematician, astronomer and philosopher whose cosmological views dominated for centuries

'In general we have to state that the heavens are spherical and move spherically; that the Earth, in figure, is sensibly spherical also when taken as a whole; in position, lies right in the middle of the heavens, like a geometrical centre; and in magnitude and distance has the ratio of a point with respect to the sphere of the fixed stars, having no local motion itself.'

From Ptolemy's *Almagest* (*c.*150 AD)

Very little is known about Ptolemy's life. He was probably of Greek ethnicity and lived in Egypt under Roman rule. His *Almagest* was one of the most influential scientific works in history and served as a basic astronomical textbook for more than a thousand years. Although there are indications that he made astronomical observations from Alexandria during the period 127 to 141 AD, Ptolemy is largely seen to have gathered earlier work by other philosophers and amended or improved it.

The *Almagest* (originally named *The Mathematical Collection*) is a synthesis of earlier Greek astronomical work and a key information source about the work of **Hipparchos**. Ptolemy extended Hipparchos's star catalogue from 850 stars to more than 1,020, and named 48 constellations visible from Egypt – their names are still used today. Ptolemy viewed the Universe as a set of nested spheres with a static Earth close to the centre, followed by the Moon, Mercury, Venus, the Sun, Mars, Jupiter and Saturn, then finally the sphere of the fixed stars. He estimated that the diameter of the sphere of the fixed stars was 20,000 times the diameter of the Earth.

His *Almagest* describes complicated mathematical constructions that accounted for the observed motion of each heavenly body fairly successfully. Although complex, it was reasonably accurate enough to explain naked-eye astronomical observations until **Tycho Brahe** made much more accurate astronomical observations 14 centuries later. To account for apparent backwards or 'retrograde' motion of the planets, due in reality to the Earth's motion around the Sun, Ptolemy made the planets move in 'eccentrics', little circles superimposed on their large circular orbits. This Ptolemaic system was dominant and widely taught for another 13 centuries until it was overthrown by the Sun-centred model of **Copernicus**.

Ptolemy's other major work was the *Geography*, an account of the geographical knowledge of the Greco-Roman world; he explained the mathematical concepts behind lines of latitude and longitude, and was aware that he only had information for about a quarter of the globe. A Latin translation of this work 13 centuries later is said to have been instrumental in persuading an Italian map-maker that Asia lay 3,000 miles west of Europe. This suggestion inspired Christopher Columbus's optimism that he could reach Asia by sailing westwards from Europe, leading to his discovery of the Americas.

Ptolemy also wrote about optics, musical intervals and astrology. His work on optics explains the principles of reflection and gives tables for the refraction angles of light rays passing from air into water. His astrological treatise, the *Tetrabiblos*, suggests a mechanism for astrological influence in terms of some kind of physical radiation from the heavens that can affect people.

GALEN
(CLAUDIUS GALENUS)
129 – C.210 AD

Roman anatomist and philosopher whose anatomical works became the dominant authority for centuries

'The fact is that those who are enslaved to their sects are not merely devoid of all sound knowledge, but they will not even stop to learn! ... Instead of admiring Nature's artistic skill – they refuse to learn; they even go so far as to scoff, and maintain that the kidneys, as well as many other things, have been made by Nature for no purpose!'

From Galen's *On the Natural Faculties* (1916)

Galen was born into a wealthy Greek family in Pergamon (now Bergama, Turkey). He travelled widely throughout the Roman empire studying medicine and he visited the medical school in Alexandria before settling in Rome around 164 AD. In Rome, he was personal physician to several emperors including Marcus Aurelius.

Galen became famous for his anatomical discoveries, made during dissection of animals including monkeys and pigs. He was the first to recognize the differences between dark blood in the veins and blood in the arteries, and he identified many muscles in the body for the first time. He also investigated the role of the spinal cord by severing it at different points in animals and recording the consequent extent of paralysis.

He realized that fluid must pass between chambers of the heart, but falsely assumed that it does so through tiny pores between them, rather than an active pumping motion by heart muscle. He thought all organs of the body consumed venous blood generated in the liver, while arterial blood came from the heart. Although many of Galen's ideas were flawed, his works were popular with Christians. He became the dominant medical authority in Europe and the Middle East for around 15 centuries, until **Harvey** and **Vesalius** radically changed the standard pictures of anatomy and physiology.

GEBER
(LATINIZED FROM ABU MUSA JĀBIR IBN HAYYĀN)
c.721 – c.815 AD

Arabic physician and alchemist who pioneered early chemical methods

'The first essential in chemistry is that thou shouldest perform practical work and conduct experiments, for he who performs not practical work nor makes experiments will never attain to the least degree of mastery. But thou, O my son, do thou experiment so that thou mayest acquire knowledge.'

Geber quoted in *Makers of Chemistry* by Eric John Holmyard (1931)

Geber was seemingly a polymath who dabbled in everything from geology and physics to medicine and alchemy. But few concrete facts are known about him. He was born in Tus, Persia (now Iran) and nearly 3,000 writings have been attributed to him, although it is unclear how many of the works were written or added to later by followers. Later writers often sought respect by attributing their works to Geber.

Geber revised the ancient Greek notion of the four elements (earth, water, fire and air) and suggested they combined to form sulphur and mercury. He believed these two could be combined in different proportions to create any metal. Iron, for instance, could be split into sulphur and mercury then recombined in different proportions to create a precious metal like gold.

Much of his writing is obtuse and mystical, but Geber's works reveal that he had mastered some chemical methods that were advanced for the time. He experimented with distillation techniques, for instance to strengthen acids, and described skilful chemical experiments in great detail. He experimented with ammonium chloride, arsenic, dyes and varnishes, and described ways to refine metals.

AL-KHWARIZMI, ABU JA'FAR MUHAMMAD IBN MUSA
*c.*780 – *c.*850 AD

Persian mathematician who introduced the modern decimal number system to the western world

'When I consider what people generally want in calculating,
I found that it always is a number'

From *Algebra* by Muhammad ibn Musa Al-Khwarizmi

Al-Khwarizmi was born into a Persian family in Chorasmia, which today belongs partly to Uzbekistan and partly to Turkmenistan. Very little is known about his life, except that he studied science and maths in Baghdad. Translated into Latin, his work played a vital role in promoting the use of the Hindu-Arabic numerals (0 to 9) and the decimal number system in Europe.

The word 'algebra' comes from the title of his chief work, *Al-jabr wa'l muqabalah*, in which he used geometric proofs to solve quadratic equations (which have the general form $ax^2 + bx + c = 0$). The word 'algorithm' also comes from the Latin form of his name. In *De numero indorum* (*On the Hindu Art of Reckoning*), Al-Khwarizmi gave a detailed account of the Hindu number system with decimal places that had been developed between the 1st and 5th centuries by mathematicians in India.

Al-Khwarizmi's work on decimal numbers had a profound effect on the pace of mathematics when Latin translations spread in the 12th century to Europe, where cumbersome Roman numerals were still the norm. As the new, more convenient decimal system was gradually adopted, it spurred progress in science, accounting and book keeping. Key to this was the use of the number zero, a concept previously unknown in the west.

Al-Khwarizmi wrote on astronomy as well as maths, compiling tables for the movement of the Sun, Moon and the five planets known at the time. He also wrote a major geographical work detailing latitudes and longitudes for 2,402 cities and landmarks.

ALHAZEN
(IBN AL-HAYTHAM)
c.965 – c.1040 AD

Persian or Arab founder of experimental optics and an early proponent of scientific rationalism

'The seeker after truth is not one who studies the writings of the ancients and, following his natural disposition, puts his trust in them, but rather the one who suspects his faith in them and questions what he gathers'

From Alhazen's *Aporias against Ptolemy* (c.1025)

Al-Haytham, known in the West as Alhazen, was born in Basra in present-day Iraq and educated there and in Baghdad. He wrote at least 100 books, the most famous being translated into Latin and eventually published in 1572 as *Opticae Thesaurus (Treasury of Optics)*.

Alhazen made detailed studies of reflection of light as well as refraction (bending) of light rays when they pass from one medium, such as air, to one of a different density, such as water. He also came up with geometric explanations for how curved mirrors create images and carried out early experiments on how white light can disperse into colours.

He argued against **Ptolemy**'s notion that our eyes send out rays to objects that we see; he recognized that the objects we look at emit light, not the eye. His research covered the eye's anatomy and rainbows, as well as the camera obscura (pin-hole camera), and he also wrote on astronomy and maths.

Some accounts of Alhazen's life claim that the sixth Fatimid caliph who ruled much of north Africa between 996 and 1021 summoned him to control flooding of the Nile. But after investigating this, Alhazen realized there was no practical solution. Fearing that the caliph would order his execution, he feigned madness and was held under house arrest for about ten years until the caliph died. He later spent many years in Islamic Spain.

NICOLAUS COPERNICUS
19 FEBRUARY 1473 – 24 MAY 1543

Polish astronomer who proposed that the Sun, rather than the Earth, lies at the centre of the Universe

'Lastly, it will be realized that the Sun occupies the middle of the Universe. All these facts are disclosed to us by the principle governing the order in which the planets follow one another, and by the harmony of the entire Universe, if only we look at the matter, as the saying goes, with both eyes.'

From Copernicus's *De Revolutionibus Orbium Coelestium* (1543)

Copernicus was born in Torun, Poland, and from 1491 studied at the University of Kracow, where he developed an interest in maths, astrology and astronomy. He held an administrative position in the church before studying canon law at the University of Bologna in Italy from 1496. In Bologna, he lived with an astronomy professor and made his first astronomical observations. He later studied medicine, probably including the use of astrology in diagnosing and treating disease, at the University of Padua.

Copernicus continued his astronomical observations while also working as a physician and in church politics. Eventually in 1510 he settled in Frombork, Poland, where he remained for the rest of his life. He wrote a short manuscript called the *Commentariolus* that introduced his heliocentric theory between 1510 and 1514. The *Commentariolus* indicates that Copernicus was troubled by flaws in traditional cosmological models. At the time, astronomy still followed principles established more than 1,800 years earlier by **Aristotle**, who held that the Earth sat at the centre of the Universe with concentric 'orbs' around it carrying the heavenly bodies in the following order: the Moon, Mercury, Venus, the Sun, Mars and Jupiter, with the most distant sphere being that of the fixed stars.

However, observers including **Ptolemy** recognized problems with this simple picture. In particular, the outer planets exhibit 'retrograde' motion – they seem sometimes to stop, move backwards and then continue forwards again. To explain that, Ptolemy introduced the concept of epicycles; as the planets orbited the Earth, they also transcribed smaller circles, which would make them appear to move backwards and forwards. The Earth was also off-centre. Ancient astronomers who subscribed to Ptolemy's theory often used it for planetary predictions without concern for whether it truly reflected the motions of the planets. Although it was complex and lacking any mechanistic explanation, it did a fairly good job of plotting the courses of the heavenly bodies.

But in the *Commentariolus*, Copernicus argued that apparent retrograde motion of the planets is much better explained by the Earth's motion around the Sun. He placed the Sun in the middle of the Universe and arranged the six planets known at the time in their correct order from the Sun. He recognized that the Moon orbits the Earth and worked out the relative scale of their orbits fairly accurately. He also showed that the seasons occur because the Earth's axis is tilted.

Over decades, Copernicus developed his theory in much greater detail in a long work, *De Revolutionibus Orbium Coelestium* (*On the Revolutions of the Heavenly Spheres*), which comprised six books. In 1539, a mathematician called Georg Joachim Rheticus (1514 – 1574) from the University of Wittenberg came to study with Copernicus and after showing him the quality of printing available in German-speaking cities, convinced Copernicus to publish *De Revolutionibus*. Rheticus oversaw the printing and a copy of the book was given to Copernicus on his deathbed.

A Lutheran minister, Andrew Osiander, also supervised the printing and added a preface to the book, which claimed that Copernicus was offering a method of calculation of planetary motions, not necessarily a realistic description of them. Possibly, he was attempting to make the work, which was dedicated to Pope Paul III, more palatable to religious opponents who believed a heliocentric system went against the Bible.

De Revolutionibus was widely read, but for about 150 years, the Copernican system was not widely accepted. Aside from religious objections, Copernicus's model did not predict planetary motions particularly well because he clung to the notion that the orbits must all be circular. This problem was not resolved until **Kepler** suggested the orbits are elliptical, an idea bolstered by **Newton**'s inverse square law of gravitation. Likewise, the invention of the telescope was a key factor in the theory's acceptance. **Galileo** used it to discover the four largest moons of Jupiter – proof that not all bodies in the solar system orbit the Earth.

Flemish anatomist who is considered the founder of modern anatomy

'Not long ago I would not have dared to turn even a hair's breadth from Galen. But it seems to me that the septum of the heart is as thick, dense and compact as the rest of the heart. I do not see, therefore, how even the smallest particle can be transferred from the right to the left ventricle through the septum.'

From Vesalius's *De Humani Corporis Fabrica* (1555 edition)

Vesalius was born in Brussels (then in the Habsburg Netherlands) to a family with a long tradition in medical careers. He studied medicine at the universities of Louvain, Paris and Padua, where he received his doctorate in 1537. He became professor of surgery and anatomy at Padua when he was only in his mid-twenties.

Vesalius was renowned for his questioning scientific method. He broke from tradition in his teaching style, carrying out dissections himself with students clustered around him rather than reading from a text book while directing an assistant to dissect an animal. He also taught using detailed drawings by himself and commissioned artists, encouraging students to check and question their findings, and even question his own.

He had access to the bodies of executed criminals for human dissection, which was starting to become acceptable practice in Italy. In doing this he realized that **Galen**, whose anatomical ideas had dominated for more than 13 centuries, had never actually dissected a human being – instead, he came to false conclusions about human anatomy after dissecting only animals, including monkeys.

In 1543, Vesalius published *De Humani Corporis Fabrica (On the Structure of the Human Body)*, a human anatomy text with unprecedented accuracy and clarity. It described and illustrated some new structures for the first time, including the thalamus that sits between the cerebral cortex and the midbrain. In the same year, he conducted a public dissection of the body of Jakob Karrer von Gebweiler, a notorious criminal from Basel, Switzerland. With the help of a surgeon, he assembled the bones and the skeleton – the oldest anatomical specimen in the world – is still on display at an anatomy and pathology museum in Basel.

The publication of *De Humani Corporis Fabrica* triggered a wave of controversy because Vesalius so radically revised the conventional wisdom laid down by Galen and **Aristotle**. He challenged Galen's view that blood can pass between the right and left chambers of the heart through tiny pores, although he stuck with the view that some sort of holes must connect the ventricles.

The human breastbone has three segments, Vesalius noted; Galen had said seven. Overall, Vesalius corrected around 200 of Galen's anatomical mistakes, including the notion that the lower human jaw has two bones rather than one. In his book, Vesalius also pointed out that both men and women have 24 ribs. Bizarrely, it was still common for people to believe that men have one less rib than women because of the Biblical tale that God took a rib from Adam to make Eve. After the book's publication, Vesalius largely gave up his research, possibly irritated by vitriolic opposition to it.

He instead became physician to the Holy Roman Emperor Charles V, and then to his son, King Philip II of Spain. Some reports suggest Vesalius was accused of heresy and was ordered to go on a holy pilgrimage to Jerusalem and Mount Sinai, or possibly he went of his own accord. In any case, during the pilgrimage, he died when his ship was wrecked off the island of Zante (often now called Zakynthos) west of Greece.

TYCHO BRAHE
(TYGE OTTESEN BRAHE)
14 DECEMBER 1546 – 24 OCTOBER 1601

Danish astronomer who became the greatest astronomical observer of the pre-telescopic age

'This new star is not located in the upper regions of the air just under the lunar orb, nor in any place closer to Earth ... but far above the sphere of the Moon in the very heavens'

Brahe's conclusions about a supernova explosion

Tycho Brahe was born into a wealthy and influential family at Knutstorp Castle (in Danish, Knudstrup borg) which is now in Sweden but was then under the Danish crown. He is usually known by his first name only (Tycho is a Latinized version of the Danish name Tyge). When he was about two years old, Tycho was effectively kidnapped by his uncle, a Danish nobleman called Jørgen Thygesen Brahe. Possibly, this was because Tycho's parents had failed to honour a promise to surrender a baby boy to Jørgen and his wife, who were childless.

Aged just 13, Tycho began studying law at the University of Copenhagen but he also developed a passion for astronomy, especially inspired by a solar eclipse on 21 August 1560 that astronomers had managed to predict in advance. He was struck by contradictions between available charts of the planets and stars, and the methods used to produce them, and determined at roughly the age of 17 that he needed to carry out a long-term, rigorous sky survey from a single location.

Between around 1576 and 1580, he built an astronomical observatory called Uranienborg ('Castle of the Sky') on the island of Hven (now a small Swedish island in the Öresund Strait) and used it to make the most accurate astronomical observations of his time. Later he was invited by the holy Roman emperor Rudolph II to Prague in 1599, and he built a new observatory near the city. During the last year of his life, his assistant at this observatory was **Kepler**, whose work deriving the laws of planetary motion relied heavily on Tycho's observations.

Tycho also dabbled with alchemy and had a life-long interest in astrology. But one of his key legacies was proving wrong **Aristotle**'s notion that the heavens are 'perfect' and unchanging. In early November 1572, he saw a 'new star' suddenly appear in the constellation Cassiopeia. (Today astronomers know that it was a very bright explosion of a white dwarf star roughly 9,000 light years away.) The following year, he published an account of it in *De nova et nullius aevi memoria prius visa stella* (*Concerning the new star, never before seen in the life or memory of anyone*). The supernova was visible to the naked eye until 1574.

Tycho's measurements proved that the new star wasn't something in the Earth's atmosphere, but occurred well beyond the Moon. That clearly contradicted the classical idea that the heavens were immutable and unchanging. In 1577, he proved that a comet was also beyond the Moon, again dealing a blow to the notion of unchanging heavens.

Kepler tried to persuade Tycho to adopt a Sun-centred model of the solar system. However, Tycho believed that if the Earth orbited the Sun each year, there should be an observable 'parallax' of the stars over six months – the change in position of the Earth would mean a change in the angular orientation of a star. This parallax does exist, but because the stars are in reality so distant, the parallax is tiny and was not detected for even the nearest stars until the 1830s (see entry for **Bessel**).

Various accounts suggest Tycho was a quarrelsome, arrogant man who easily got into fights. In 1566, aged 20, he lost part of his nose in a duel against another Danish nobleman over a mathematical dispute, and he wore a false metal nose for the rest of his life. But he also enjoyed large parties in his castle and had eight children with his common-law wife. He contracted a bladder or kidney disease after going to a banquet in Prague and died within two weeks. His unprecedentedly accurate catalogue of nearly 800 stars was published after his death.

JOHANNES KEPLER
27 December 1571 – 15 November 1630

*German mathematician and astronomer whose laws of
planetary motion revolutionized science*

'We do not ask for what useful purpose the birds do sing, for song is
their pleasure since they were created for singing. Similarly, we ought
not to ask why the human mind troubles to fathom the secrets of the
heavens. The diversity of the phenomena of nature is so great and
the treasures hidden in the heavens so rich precisely in order that the
human mind shall never be lacking in fresh nourishment.'

From Kepler's *Mysterium Cosmographicum* (1596)

Kepler was born in Weil-der-Stadt, near
Stuttgart. From 1589, he studied at the University of
Tübingen and in 1593, he was appointed to a teaching
post in Graz. He corresponded with the astronomer
Tycho Brahe, who was then in Prague, and later
became his assistant until the latter's death in 1601.
After that, Kepler was appointed as Tycho's successor
as Imperial Mathematician in Prague, the most
prestigious appointment in mathematics in Europe. He later worked at
Linz and at Sagan in Silesia.

Kepler developed a strong interest in astronomy in childhood.
He was inspired by his mother taking him, aged just six, to see a bright
comet in the night sky in 1577. Three years later, he recalled being taken
outside to see an eclipse of the Moon and that the Moon's colour turned
a distinct red.

While working at Graz, Kepler published his famous work *Mysterium
Cosmographicum* (*The Cosmographic Mystery*) in 1596. In it, he argued
that the distances of the planets from the Sun in the Copernican system
were determined by the five regular solids (with 4, 6, 8, 12 and 20 sides),
if a planet's orbit was circumscribed around one solid and inscribed in
another. This reflected Kepler's lifelong wish to establish a divine harmony
in the Universe. It was also a surprisingly accurate model for some of the
planetary orbits, given that it had no genuine physical basis.

In 1600, Kepler became an assistant to Tycho, who was then Imperial Mathematician in Prague. Tycho asked him to calculate new orbits for the planets from his unprecedentedly accurate observations, which Tycho left to Kepler following his death in 1601. Over several years, Kepler came to conclusions that would dramatically alter astronomers' picture of the laws of nature. Since ancient times, dogma dictated that planetary orbits must be circular. Kepler showed convincingly that astronomical observations made much more sense if the planets followed ellipses in their course around the Sun (he had already accepted the heliocentric system of **Copernicus**).

Kepler formulated three key laws of planetary motion. The first law stated that the orbit of a planet around the Sun is an ellipse with the Sun's centre of mass at one focus, and his second law pointed out that a line joining a planet and the Sun sweeps out equal areas in equal intervals of time (this reflects the fact that the Sun's gravity accelerates a planet's speed when it is close to the Sun). Kepler's third law states that the squares of the periods of the planets are proportional to the cubes of their semi-major axes (effectively, the longest 'radius' of the ellipse).

In addition to these major breakthroughs, Kepler explained how the tides were influenced by the Moon. He showed, as Tycho had done previously, that a 'new star' in 1604 (now known to have been a supernova explosion) had features that indicated it must lie well beyond the solar system, again dealing a blow to the ancient idea of the unchanging distant sphere of the fixed stars. He also investigated optics and invented an improved design for the refracting telescope, and he wrote a description of a bright comet of 1607, later known as **Halley**'s Comet after Halley correctly predicted its future return date.

Kepler corresponded with **Galileo** following Galileo's discovery of the four moons of Jupiter in 1610, and endorsed the discoveries with enthusiasm. His work is often seen as marking the transition point when scientific mysticism gave way to rigorous, rational science in the modern sense. But Kepler's most important legacy was his laws of planetary motion, which felled erroneous views that had gone unquestioned for many centuries. They were instrumental in laying the foundations for **Newton**'s laws of gravity.

FRANCIS BACON
22 January 1561 – 9 April 1626

*English philosopher and statesman who pioneered the
modern scientific method*

'If a man will begin with certainties, he shall end in doubts; but if he
will be content to begin with doubts, he shall end in certainties'

From Bacon's *The Advancement of Learning* (1605)

**Francis Bacon studied at Cambridge
University** and Gray's Inn, London. He then
began a political career, becoming a member of
parliament in 1584. He was knighted in 1603 and
later became Lord High Chancellor in 1618, but he
was banished from office in disgrace in 1621 and
imprisoned in the Tower of London for four days
after admitting taking bribes. He was deep in debt
when he died of pneumonia, supposedly contracted
while conducting meat-freezing experiments.

Bacon's key contribution to science was to promote rational, evidence-
based argument. Many scientific ideas of the period still harkened back
to **Aristotle** and his followers. Although **Copernicus**'s work had already
discredited the notion of an Earth-centred Universe, many philosophers still
believed that scientific truth could emerge from simple discussion of ideas.

In his *Novum Organum Scientiarum* (*New Instrument of Science*, 1620),
Bacon argued that experimental evidence from the real world was required.
His *New Atlantis* (1624) was an influential utopian novel, in which he
described a scientific academy he called 'Salomon's House' where trained
scientists sought out data, conducted experiments, then applied the
knowledge they gained for practical benefit to society – essentially the
concept of today's applied science universities.

Bacon stressed the importance of facts and experience, and that truth
should not be passively accepted from authority. He criticized Renaissance
alchemy and astrology for their lack of evidence base, and his views were
influential on **Hooke**, **Boyle** and other scientists who later founded the
Royal Society as a research institution in 1660.

WILLIAM HARVEY
1 April 1578 – 3 June 1657

English physician who gave the first complete description of the circulation of the blood

'[The heart] is the household divinity which, discharging its function, nourishes, cherishes, quickens the whole body, and is indeed the foundation of life, the source of all action'

From Harvey's *Exercitatio Anatomica* (1628)

Harvey was born in Folkestone, Kent. He was educated at King's College in Canterbury and at Cambridge University before studying medicine at the University of Padua in Italy. He later became a fellow of the Royal College of Physicians and physician to St Bartholomew's Hospital. He also served as physician to British kings James I and Charles I, who both encouraged Harvey's research.

By 1616, Harvey had carried out dissection experiments on 80 species of animal. At the time, anatomical errors by **Galen** had remained virtually unchallenged for 1,400 years. Galen thought blood formed in the liver and passed between the chambers of the heart through tiny pores rather than being actively pumped. Harvey showed that in fact the heart is a muscle that pumps blood by contracting and that it pumps so much blood in an hour that the blood couldn't possibly be formed and broken down quickly enough. Instead, he proposed that the same blood circulates from the heart through the arteries, then through veins back to the heart.

Harvey could see no visible connections between the arteries and the veins. But he knew they both subdivided into ever finer vessels and argued that connections between them must occur where they had become too small to see. **Malpighi** confirmed this in the early 1660s. Harvey also carried out early experiments on the development of chicks in eggs as well as deer embryos. His observations convinced him that embryos develop by 'epigenesis', the gradual addition of parts, as opposed to the popular notion of 'preformationism' – that animals develop from fully formed but miniature versions of themselves.

GALILEO GALILEI
15 February 1564 – 8 January 1642

Italian astronomer and natural philosopher who formulated principles of motion, backed the Sun-centred model of the solar system and pioneered early telescope observations

'Eppur si muove' ('And yet it does move')

Alleged comment uttered by Galileo after he was forced to renounce the notion that the Earth moves round the Sun

Born in Pisa, Galileo studied medicine at the University of Pisa from 1581 to fulfil his father's wishes, but he took the opportunity there to pursue his interests in maths and natural philosophy. Having worn his father down, he continued private studies of the works of famous natural philosophers including **Euclid** and **Archimedes** and gave up his medical course, leaving without completing his degree. After several teaching jobs, he was appointed professor of mathematics at the University of Pisa in 1589.

Following the death of his father, Galileo was keen to support his family and he successfully secured a more lucrative post as professor of mathematics at the University of Padua in 1592. He remained there for 18 years and taught Euclidean geometry and astronomy to medical students, so that they could apply astrology in their medical practice.

Early in his career, Galileo did not publicly support **Copernicus**'s Sun-centred view of the solar system. However, like **Tycho Brahe**, he did argue that a 'new star' in 1604 (now known to have been a supernova explosion) must lie well beyond the solar system. This dealt a blow to **Aristotle**'s still-prevalent theory that all changes in the heavens had to occur within the realm of the planets, the 'sphere of the fixed stars' being permanent.

By around 1604, Galileo had also formulated his theory of motion, showing that a projectile follows a parabolic path. In experiments with pendulums and balls rolling down inclined planes, he demonstrated that the speed of a falling body does not depend on its mass and that, falling from rest, the distance it travels is proportional to its acceleration and the square of the time elapsed. However, he did not publish these results until more than three decades later. He also clung incorrectly to the ancient notion that a body with no force acting on it would naturally follow a circular path.

Galileo is often credited with having invented the telescope, but he probably was not the first to build one. He did, however, skilfully construct a series of simple refracting telescopes, achieving a magnification of about a factor of nine by 1609. He was also the first person to enthusiastically throw himself into detailed telescopic observations of the night sky and made some dramatic discoveries that he outlined in his publication *Siderius Nunicus* (*Starry Messenger*) of 1610. In it, he claimed to have found evidence for mountains on the Moon and he argued that the Milky Way is brimming with countless stars.

Perhaps most sensationally of all, he had by now discovered four moons circling Jupiter (now known as the Galilean moons Io, Europa, Ganymede and Callisto). This established the telescope's astonishing power to reveal objects invisible to the naked eye, and crucially provided incontrovertible evidence that celestial bodies can orbit a world other than the Earth. That left **Ptolemy**'s geocentric theory – in which all bodies orbited the Earth – in serious trouble.

Around the same time, Galileo's observations of Venus showed that the planet undergoes phases, and that it therefore must orbit the Sun closer than the Earth. This was not solid proof for the Copernican heliocentric model exclusively, but Galileo found it a persuasive factor. Although he tried to avoid controversy with the Catholic Church, which viewed the heliocentric model as heretical, he eventually became drawn into the debate and argued that it was important to interpret the Bible in light of scientifically established fact, and that the heliocentric system reflected physical reality, rather than simply being a useful model for calculations. To quell the controversy, in 1616 Pope Paul V instructed Galileo to cease supporting Copernicus publically.

The fuss died down, however, and much later Galileo was granted permission to publish *Dialogue Concerning the Two Chief Systems of the World* (1632). This pitted the Aristotelian model against the Copernican one as a dialogue between two characters, but it portrayed the Aristotelian supporter as a simpleton, possibly even parodying the contemporary Pope himself, Urban VIII. Galileo was summoned to Rome and forced under threat of torture to renounce the heliocentric model. He escaped violent punishment but was confined under house arrest until his death.

RENÉ DESCARTES
31 MARCH 1596 – 11 FEBRUARY 1650

French philosopher and mathematician who introduced the Cartesian coordinate system

'Cogito ergo sum' ('I think, therefore I am')

From Descartes's *Principles of Philosophy* (1644)

Descartes was born in La Haye en Touraine (now named Descartes) in central France and studied law at the University of Poitiers. He signed up as a volunteer for the Dutch army and after travelling in Europe for about 11 years, he lived in the Netherlands for most of the rest of his life.

Descartes wrote extensively on philosophy. Suspicious of commonly accepted knowledge because of the subjective nature of the senses, he attempted to rebuild knowledge on a more objective footing. His best-known saying, 'I think, therefore I am', reflects his view that if someone is capable of wondering whether or not they really exist, that in itself is proof of their existence.

In mathematics, Descartes's contributions were ground-breaking. In algebra, he was the first to use early letters of the alphabet for constants and late letters for variables, a convention that persists today. Most importantly, he invented the Cartesian coordinate system. Seeing a fly buzzing around his bedroom, he realized he could define its position precisely relative to three axes each at right angles to the other, with an 'origin' where they intersect.

He used positive and negative numbers to denote distances from the origin in opposite directions along each axis and showed how curves could be represented in this geometry. For example, in a two-dimensional plane with an x and y axis, $y = 3x^2 - 5$ represents a parabola. This fusion of algebra and geometry paved the way for **Newton** and **Leibniz** to develop calculus, the maths of changing quantities, which proved to be an excellent tool for describing nature's basic laws.

OTTO VON GUERICKE
20 November 1602 – 11 May 1686

German scientist, inventor and politician who pioneered vacuum experiments

'Since everything is in Space, and since it provides a place for things to be and to continue to be, what could Space itself possibly be? Is it perhaps some heavenly material of the nature of fire – solid (according to the Aristotelians) or fluid ... Or is Space empty of all material? Is it indeed the perpetually denied vacuum?'

From a translation of Guericke's works (*Thinking About Nothing* by Thomas Conlon, 2011)

Born in Magdeburg, von Guericke studied at the University of Leipzig from 1617 and later at the universities of Jena and Leiden, where he took courses on maths, physics and engineering. He then travelled in France and England. He returned to Magdeburg in 1627, and when the city was burned by Imperial forces in 1631, his family managed to escape but lost all their possessions. He later worked as an engineer during the regeneration of the town, where he served as mayor from 1646 for more than 30 years.

Von Guericke became interested in philosophical debates about the vacuum. **Aristotle** had reasoned that as air becomes less dense, objects can move through it more quickly and that in a vacuum, objects would move with infinite speed. Deciding that this was absurd, Aristotle concluded that a vacuum cannot exist.

To test this, von Guericke invented the first air pump in 1650 and embarked on a series of elaborate and expensive experiments with evacuated vessels, in which he showed that a ringing bell is silenced and candles can't burn. He also demonstrated that air pressure prevented two teams of eight or more horses pulling apart two hemispheres with the air between them evacuated.

BLAISE PASCAL
19 June 1623 – 19 August 1662

French mathematician, physicist and religious philosopher who developed probability theory

'There are then two kinds of intellect: the one able to penetrate acutely and deeply into the conclusions of given premises, and this is the precise intellect; the other able to comprehend a great number of premises without confusing them, and this is the mathematical intellect. The one has force and exactness, the other comprehension.'

From Pascal's *Pensées* (1669)

Born in Clermont-Ferrand, Pascal was a child prodigy who was educated by his father, a mathematician and local administrator. While still a teenager, he designed some simple calculating machines hoping to assist his father's laborious accounting work, and this led to him inventing the mechanical calculator. He eventually built 20 machines in total, continually improving their design.

Aged just 16, Pascal proved what's today known as Pascal's theorem, stating that if a hexagon is inscribed in any circle, or more generally any section of a cone, and pairs of opposite sides are extended until they meet, the three intersection points will lie on a straight line. (This is always true for an irregular hexagon with sides of varying length.) His work was so impressive that **Descartes** refused to acknowledge that it could be the work of a youth.

In 1653, Pascal published a work describing what's now known as Pascal's triangle, often written as a triangular array of numbers in which each number is the sum of the two numbers above it. The sequential rows of a simple Pascal triangle relate the number of coin tosses to the number of possible outcomes, for instance, and the concept proved useful in several aspects of probability theory, which Pascal developed in collaboration with the French mathematician Pierre de Fermat. Reading of their work, **Huygens** wrote the first book on probability theory, which was later built on by **Laplace** and others.

In physics, Pascal studied the principles of hydraulic fluids. His inventions include the hydraulic press, which uses hydraulic pressure to amplify force, and the syringe. He showed that pressure applied to a liquid in a confined space is exerted in all directions through the liquid, not just the direction in which it is applied.

By 1646, Pascal had heard about the experiments of Italian physicist and mathematician Evangelista Torricelli, who invented the barometer, and he was intrigued to replicate them. In one experiment, he placed a tube of mercury (then referred to as 'quick silver') in a bowl of mercury, the tube open at the bottom, and noted that some of the mercury nonetheless stayed in the tube with a gap above it. Pascal questioned what force kept some mercury up in the tube and what filled the gap above it.

At the time, it was still common to hold **Aristotle**'s view that it was impossible for a vacuum to exist. Aristotle had reasoned that as air becomes less dense, objects can move through it more quickly and that in a vacuum, objects would move with infinite speed – an absurd notion, he decided, which meant that a vacuum was impossible. But Pascal demonstrated the rules describing the support of various liquids by pressure and gave persuasive proof that a vacuum, or near vacuum, held up the column of liquid in a barometer tube. He also demonstrated that air pressure near the Earth's surface decreases with altitude by carrying a barometer to the top of volcanic lava dome and a high church tower.

Following a religious experience in 1654, Pascal largely gave up maths and turned to writing. He was converted to a sect of Catholicism known as Jansenism and wrote prolifically on theology. Having suffered ill-health for most of his adult life, his condition dramatically worsened in 1662, and he died in Paris aged only 39 after suffering extreme convulsions. A post-mortem revealed that he had a disease of the abdomen and brain damage, although the cause of his condition in modern terms is not clear. The standard unit of pressure or stress is named the pascal in his honour, as is the Pascal computer programming language, developed from the late 1960s.

ROBERT HOOKE
18 July 1635 – 3 March 1703

British experimental scientist whose wide interests spanned physics, astronomy, chemistry, biology, geology and technology

'It is the great prerogative of Mankind above other Creatures, that we are not only able to behold the works of Nature, or barely to sustein our lives by them, but we have also the power of considering, comparing, altering, assisting, and improving them to various uses ... The truth is, the Science of Nature has been already too long made only a work of the Brain and the Fancy: it is now high time that it should return to the plainness and soundness of Observations on material and obvious things.'

From Hooke's *Micrographia* (1665)

Hooke was born at Freshwater on the Isle of Wight. He went to Westminster School at the age of 13 and from there went to Oxford, where he impressed scientists with his talent for designing and building experiments. He was appointed as an assistant to **Boyle** before, in 1662, becoming curator of experiments at the newly formed Royal Society of London. He later became professor at Gresham College, London.

For his work on biology, Hooke designed the compound microscope – one of the best of its time – and used it for demonstrations at the Royal Society's weekly meetings. With it he observed organisms as diverse as sponges, insects and bryozoans (tiny aquatic invertebrates). The Royal Society published descriptions of them in Hooke's beautifully illustrated *Micrographia* (1665), often considered the first scientific bestseller. The diarist Samuel Pepys called it 'the most ingenious book that I ever read in my life'.

In microscope studies of cork, Hooke noted structures that reminded him of the cells of a monastery, and this led him to coin the biological term 'cell'. Through microscope examinations of fossils, he recognized that minerals from water can penetrate dead wood, fossilizing it into petrified wood. He also understood that inland and mountain fossil beds formed in much earlier times when the land was submerged in water, before the land rose up above the water's surface.

Hooke noted that many fossils represented organisms no longer known to science: 'There have been many other Species of Creatures in former Ages, of which we can find none at present; and that 'tis not unlikely also but that there may be divers new kinds now, which have not been from the beginning,' he wrote in his *Discourse of Earthquakes*, published after his death in 1705, but probably outlined by Hooke in lectures to the Royal Society in the year 1668 or earlier. He had grasped, around 250 years before **Darwin**'s time, that fossils document species that have both appeared or disappeared during Earth's history.

He was also a prolific inventor, designing the universal joint (which connects two sections of rigid rod and allows them to bend at any angle to each other) and the iris diaphragm used in cameras. He improved many meteorological instruments and following the Great Fire of London in 1666, he played a leading role in the rebuilding effort as chief surveyor.

But he's probably best remembered for the relatively mundane Hooke's law, an equation describing elasticity, because it crops up commonly in modern school textbooks. He showed that the extension of a spring or material is directly proportional to the load applied to it, so long as the load doesn't stretch the material beyond its elastic limit. Hooke had a reputation for being irascible and clashing over priority disputes with his intellectual rivals, most famously **Newton**, who probably did much to obscure Hooke's contributions.

ROBERT BOYLE
25 January 1627 – 31 December 1691

Irish–English physicist, philosopher and inventor who is also considered one of the founders of modern chemistry

'If we consider God as the author of the Universe, and the free establisher of the laws of motion, whose general concourse is necessary to the conservation and efficacy of every particular physical agent, we cannot but acknowledge, that, by withholding his concourse, or by changing these laws of motion, which depend perfectly upon his will, he may invalidate most, if not all the axioms and theorems of natural philosophies'

From Boyle's *Reason and Religion* (1675)

Boyle was born in Lismore Castle in County Waterford, Ireland, to an English father. From the age of eight, he spent three years at Eton then travelled in Europe with a French tutor, in 1641 visiting Italy where they studied the works of **Galileo**. Boyle returned to England in 1644, by which time his wealthy father had died and left him a substantial estate, leaving him financially independent enough to devote his life to study.

Boyle settled at Stalbridge in Dorset, where he first wrote on moral and literary topics but from 1649 his interest turned to experimental science, which dominated his subsequent career. He began experiments in chemistry and alchemy, and used a microscope to observe the minute structure of living things.

In late 1655 or early 1656, Boyle moved to Oxford to join a group of natural philosophers who shared enthusiasm for acquiring knowledge through experimental investigation. They also met frequently in London, often at Gresham College. This group is often seen as a precursor to the Royal Society, founded in 1660. During this period Boyle studied the work of major European philosophers including **Descartes**.

He also met **Hooke** and employed him as an experimental assistant. Hooke built air pumps for vacuum chambers in which Boyle demonstrated the principle that made him famous, Boyle's law. This states that the pressure of a confined gas is inversely proportional to its volume, if the temperature remains the same. Since air was compressible, it presumably was composed of discrete particles that got closer together on compression, so this work supported **Dalton**'s 'atomist' theory, rather than the notion that matter can be broken down indefinitely.

Boyle was also first to demonstrate in a partial vacuum that all objects fall at the same rate, as Galileo's work suggested, and he showed that a vacuum can transmit electrical forces but not sound. In the 1600s, he wrote an extraordinarily visionary 'wishlist' of 24 feats that he believed science could or should achieve. It anticipates aviation, organ transplants, technologies for accurate map-making, commercial agriculture and psychotropic drugs, and 'the recovery of youth, or at least some of the marks of it, as new teeth, new hair coloured as in youth'. The vast majority of his predictions did eventually come about.

In 1668, Boyle left Oxford for London, where he lived with his sister for 20 years. During his lifetime, he published more than 40 books. Boyle also wrote extensively on theology. As the quote above suggests, he believed that God was free to create any kind of Universe and that even after creating it, God remained free to intervene with the laws of nature and change them.

MARCELLO MALPIGHI
10 March 1628 – 29 November 1694

Italian anatomist who pioneered early microscopy

'I could clearly see that the blood is divided and flows through tortuous vessels and that it is not poured out into spaces, but is always driven through tubules and distributed by the manifold bendings of the vessels ... If Nature once circulates the blood within vessels and combines their ends in a network, it is probable that they are joined by anastomosis at other times too.'

Malpighi quoted in *Marcello Malpighi and the Evolution of Embryology* (1966, vol 1)

Malpighi was born in Crevalcore, near Bologna. He studied medicine, anatomy and philosophy at the University of Bologna, where he earned his medical degree in 1653. He then held teaching posts and professorships at various Italian universities, and from 1663 worked as a physician at Bologna while conducting experiments in his spare time. In 1691, he moved to Rome as physician to Pope Innocent XII.

In microscopy experiments on frogs, Malpighi showed that blood flows through a complex network of vessels in the lungs, clarifying a key step in respiration by explaining how air enters the bloodstream to reach all parts of the body. In bats, he discovered the tiny blood vessels later named capillaries and showed they connect the smallest visible veins with the smallest visible arteries, a key link enabling the blood circulation envisioned by **Harvey**.

Malpighi was also among the first scientists to use a microscope to examine chick embryos at very early stages, observing the formation of structures that would develop into the heart and blood vessels. Many anatomical structures are named after him, including the Malpighian layer of the skin.

GOTTFRIED WILHELM LEIBNIZ
1 July 1646 – 14 November 1716

German mathematician and philosopher who developed calculus

'In whatever manner God created the world, it would always have been regular and in a certain general order. God, however, has chosen the most perfect, that is to say, the one which is at the same time the simplest in hypothesis and the richest in phenomena.'

From Leibniz's *Discours de Métaphysique* (1686)

Born in Leipzig, Saxony, Leibniz attended the University of Leipzig from the age of 14 and earned a doctorate in law aged 20 from the University of Altdorf. He then became a legal adviser to the Elector of Mainz. In his spare time, he studied the works of **Newton**, **Descartes**, **Pascal** and **Boyle**. In 1672, he was sent on a diplomatic mission to Paris, where he spent four years before becoming librarian to the Duke of Brunswick in Hanover, a post he held till he died.

Under the tutorship of **Huygens** in Paris, Leibniz developed calculus, one of the greatest mathematical developments of all time. Calculus, the study of rates of change, opened up the way to solving diverse problems in science, engineering and economics for which algebraic formulas were insufficient. Leibniz later became embroiled in a priority dispute with Newton over the invention of calculus. The Royal Society ruled for Newton in 1711, but it is the superior notation Leibniz invented that is still in use today.

Leibniz also developed various technological ideas, designing calculating machines, improved balance wheels for watches, a submarine and an aneroid barometer. Throughout his life, he wrote and translated vast amounts of philosophy, literature and poetry. In philosophy, he famously tried to demonstrate that we live in the best of all possible worlds, a stance that Voltaire ridiculed in his satirical novel *Candide* (1759).

JOHN FLAMSTEED
19 August 1646 – 31 December 1719

English astronomer who catalogued more than 3,000 stars as the first Astronomer Royal

'It only remains that I give you the answer I would make to our suggesting friend [Halley], when he asks me why I do not print my observations? 'Tis first I do not find myself under any obligations to receive instructions what to do, or be governed by him and his associates ... Secondly, I would not thrust such an incomplete catalogue on the world as he has done from St Helena ... He has more of mine in his hands already, than he will either own or restore; and I have no esteem of a man who has lost his reputation, both for skill, candour, and ingenuity, by silly tricks, ingratitude and foolish prate; and that I value not all or any of the shame of him and his infidel companions.'

In a letter to **Newton**, Flamsteed vents his anger at **Halley**

Born in Denby, Derbyshire, Flamsteed developed an early interest in maths and astronomy. He suffered from a chronic rheumatic condition that made his father decide not to send him to university, but from 1662 he studied astronomy on his own. In 1665, he wrote a paper on astronomy that outlined the design and construction of an astronomer's quadrant as well as a way of measuring the Sun's distance from the Earth. In 1670, Flamsteed visited Cambridge University and enrolled as an undergraduate, and while he seemingly didn't move there to study full time, he was there for at least part of 1674, when he attended **Newton**'s lectures.

One pressing problem of his age was determining longitude at sea. Latitude is easy to measure at any location from the maximum altitude of the Sun at midday, for instance, or from the elevation of the star Polaris directly above the Earth's north pole. With no accurate way of estimating longitude, however, ships could run wildly off course, sometimes leading to shipwrecks, or to crews succumbing to scurvy or starvation. One way to determine longitude was to keep a standard-time clock onboard ship and note the difference between standard and local time, measured by the Sun or stars; yet no pendulum clock could keep time accurately on a rolling ship.

An alternative proposal was to determine standard time by tracking the Moon's motion against background stars. However, Flamsteed pointed out that star maps of the 1600s were not accurate enough to allow this. Hearing of this, King Charles II appointed him first Astronomer Royal, charged with constructing accurate star tables, for which he provided him with a new observatory at Greenwich. In 1675, Flamsteed himself laid the foundation stone for the Royal Greenwich Observatory, designed by Christopher Wren.

Over the next few decades, Flamsteed compiled the most extensive and accurate star catalogue of its time. It contained more than 3,000 stars and tripled the number of stars in **Tycho Brahe**'s catalogue. But by many accounts, he was an uncompromising and confrontational man who easily fell out with his peers, possibly due to professional jealousy. In the latter part of his life, he became embroiled in a bitter dispute with Newton and **Halley**, who became frustrated at Flamsteed's pernickety attitude to his observations. He had spent nearly 30 years at the Royal Observatory yet had not published his star charts, insisting that he wait until his observations were complete.

Through the Royal Society, Newton led a campaign to pressure him into publication, while in 1704 Prince George of Denmark offered to fund the printing. Under duress, Flamsteed handed over the incomplete observations, which were edited by Halley, and 400 copies were printed in 1712. Later, Flamsteed got hold of 300 copies and burnt them. In his diary, he wrote that this was 'a sacrifice of them to Heavenly Truth; as I should do of all the rest of my editor's pains of the like nature, if the Author of Truth should put them into my power'.

Flamsteed's astronomical measurements set new standards in astronomy and allowed more stringent tests of scientific theories such as the laws of gravitation. However, the problem of longitude determination was eventually solved instead by English clockmaker John Harrison, who in the 1750s invented the marine chronometer, a clock that could accurately keep time on a ship.

Among Flamsteed's other contributions to astronomy, he accurately predicted two eclipses of the Sun in 1666 and 1668, and from 1690, he observed the planet Uranus. He was probably among the first people to do so, although he did not recognize that it was a planet and catalogued it as a star. In 1725, Flamsteed's own version of his star catalogue, *Historia Coelestis Britannica*, was published posthumously. Halley, who Flamsteed despised, was appointed Astronomer Royal as his successor.

ANTONIE PHILIPS VAN LEEUWENHOEK
24 October 1632 – 26 August 1723

Dutch scientist who pioneered microbiology

'My work, which I've done for a long time, was not pursued in order to gain the praise I now enjoy, but chiefly from a craving after knowledge, which I notice resides in me more than in most other men. And therewithal, whenever I found out anything remarkable, I have thought it my duty to put down my discovery on paper, so that all ingenious people might be informed thereof.'

From a letter by Leeuwenhoek (12 June 1716)

Born into a trading family in Delft, Holland, Leeuwenhoek had no formal science training but despite that became one of the greatest scientists of his age. After schooling he became a fabric merchant but his main hobby, which he began at some point before 1668, was grinding lenses and glass blowing to make single-lens microscopes.

Leeuwenhoek's tiny but exquisitely shaped lenses, most less than 3 mm wide, were capable of magnifying up to 200 times or more, far greater than his contemporaries could achieve. He made observations with his microscopes in painstaking detail and commissioned an artist to illustrate his findings accurately. Nothing lay beyond his curiosity – he studied everything from blood cells to ditch water and spiders' webs as well as plaque from between his own teeth.

From 1673, Leeuwenhoek wrote regular letters about his observations to the Royal Society in London. He was among the first to observe bacteria (which he called animalcules). He described yeast cells and discovered free-living protists for the first time, and observed parasites on fleas, blood cells and microscopic nematodes. He was first to record microscopic observations of muscle fibres and studied blood flow in capillaries. He also discovered sperm cells in human and animal semen, and gradually came to the conclusion that they penetrate egg cells during fertilization.

GIOVANNI DOMENICO CASSINI
8 June 1625 – 14 September 1712

*Italian astronomer who made the first accurate
measurements of the size of the solar system*

'The breadth of the ring was divided by a dark line into two equal
parts, of which the interior and nearer one to the globe was very
bright, and the exterior part slightly dark. There was about the
same difference between the colours of these two parts that there is
between dull silver and burnished silver.'

From Cassini's paper '*The Discovery of the Division in Saturn's Ring*'
(in *Mémoires de l'Académie Royale des Sciences*, 1730)

Cassini was born in Perinaldo in the Republic of Genoa near
Nice, France, and taught astronomy at the University of Bologna from
1650. In 1669, he moved to France and was funded by King Louis XIV
to set up the Paris Observatory, which opened in 1671. Between 1671 and
1684, Cassini discovered four of Saturn's five biggest moons, which are
now called Tethys, Dione, Rhea and Iapetus. (Saturn's largest moon, Titan,
had been discovered in 1655 by **Huygens**.)

He also discovered the 'Cassini division', a 4,800 km (2,980 mile) gap
between Saturn's two brightest rings, which he correctly suspected are
made up of myriad tiny particles. In 1672, Cassini made the first scientific
and accurate attempt to measure the distance to Mars. His colleague Jean
Richer travelled to Cayenne, French Guiana, while Cassini stayed in Paris
and they made simultaneous measurements of Mars's location in the sky.

By triangulation, Cassini was able to calculate the distance to Mars
with 93 per cent accuracy. For the first time, this allowed an estimate of the
overall size of the solar system, since the relative distances of the Sun and
planets were already known from **Kepler**'s third law. Cassini's estimate
of the Sun's distance – 87 million miles – was the first estimate that was
roughly right.

CHRISTIAAN HUYGENS
14 April 1629 – 8 July 1695

Dutch mathematician, astronomer and physicist who proposed the wave theory of light and invented the pendulum clock

'Light comes from the luminous body to our eyes by some movement ... light takes time for its passage ... it will follow that this movement, impressed on the intervening matter, is successive; and consequently it spreads, as sound does, by spherical surfaces and waves: for I call them waves from their resemblance to those which are seen to be formed in water when a stone is thrown into it, and which present a successive spreading as circles.'

From Huygens's *Treatise on Light* (1690)

Born in The Hague, Huygens studied law and mathematics at the University of Leiden and the College of Breda. In the 1650s, he did a range of work in science and maths, constructing telescopes, microscopes and clocks, and after diplomatic service he moved to Paris in 1666, where he worked for the French Academy of Sciences under the patronage of Louis XIV. He later carried out astronomical research at the Paris Observatory.

Encouraged by **Pascal**, Huygens wrote the first book on probability theory *De ratiociniis in ludo aleae* (*On Reasoning in Games of Chance*, 1657). Around this time he also derived the formula for the strength of the centripetal force, for instance the force exerted by a tennis ball whirled around on a string. He showed that it is equal to mv^2/r, where m is the ball's mass, v is its velocity and r is the radius of the string.

Working with a Dutch clock-maker, Huygens invented the pendulum clock, patented in 1657. But he is probably best remembered for arguing the case that light consists of waves. He assumed that light can only be transmitted through some kind of imperceptible medium of particles, which pass on the disturbance to their neighbour in a similar way to sound waves. Each point on a wavefront emitted secondary wavelets, which joined to create an envelope that represented a new wavefront.

He incorrectly assumed that light waves are longitudinal like sound, rather than transverse, but he used his theory to successfully explain reflection and refraction of light, as well as the slowing of light when it moves from one medium into a denser one.

Huygens's wave theory of light was vindicated by **Young**'s experiments on light interference. Later, **Newton** argued that light consists of discrete particles, confirmed by **Einstein**'s discovery of the photoelectric effect in the early 1900s. However, quantum theory reconciled these contradictions by introducing the peculiar notion of wave–particle duality – light behaves as both particles and waves depending on the way it is observed.

In 1673, Huygens published his major work (*Horologium Oscillatorium*) on analysis of pendulums, clocks, mechanics and geometry. Other scientists had already noted that a swinging pendulum does not always have exactly the same period – wider swings take slightly longer than gentle ones. Using mathematics analogous to an early application of calculus, Huygens showed that to maintain a constant period regardless of amplitude, a pendulum bob would have to move in a curve called a cycloid, not the arc of a circle.

He also designed a pendulum support that would force the bob to follow a cycloidal path, which the clock-maker Salomon Coster used in his construction of a pendulum clock, although in reality this solution was less practical than simply limiting the pendulum's swing amplitude to just a few degrees. In all his work on mechanics, Huygens did not accept Newton's view of gravity as a force, disliking the way it conjured up the notion of inexplicable 'action at a distance'.

Huygens also discovered the phenomenon of coupled oscillations. He noticed that two of his pendulum clocks placed on the same support gradually altered their relative phase until they were synchronized, swinging in opposite directions. He referred to it as 'an odd kind of sympathy', and he went on to conclude correctly that the effect was due to 'imperceptible movements' in the clocks' support.

In astronomy, Huygens designed his own refracting telescopes and in 1655 after observing Saturn he proposed that it has a giant ring system. In the same year, he also discovered Saturn's biggest moon, Titan. He had forward-thinking views on extraterrestrial life, arguing that other planets could support life if they had liquid water, although he recognized that Venus would be too hot for that and Jupiter would be too cold. The European Space Agency's Huygens probe, which landed on Titan in 2005 after separating from NASA's Cassini orbiter, was named in his honour.

ISAAC NEWTON
4 January 1643 – 31 March 1727

English giant of science who formulated laws of motion and gravity

'I do not know what I may appear to the world, but to myself I seem to have been only like a boy playing on the sea-shore, and diverting myself in now and then finding a smoother pebble or a prettier shell than ordinary, whilst the great ocean of truth lay all undiscovered before me'

Newton quoted in *Memoirs of the Life, Writings, and Discoveries of Sir Isaac Newton* by David Brewster (1855)

Newton was born prematurely in Lincolnshire three months after the death of his father. As a teenager, he attended school in Grantham then Cambridge University, from 1661. At the time, **Aristotle**'s theories still dominated education but Newton was much more interested in the modern theories of astronomers and philosophers like **Galileo**, **Kepler** and **Descartes**. Afterwards, he rigorously developed theories at home during a plague epidemic before returning to Cambridge in 1667 as a fellow of Trinity College. He was appointed Lucasian Professor of Mathematics there in 1669.

His book *Philosophiæ Naturalis Principia Mathematica* (often just called the *Principia*), published in 1687 and still on sale as a paperback today, outlines Newton's three laws that describe the relationship between a force acting on a body and the body's motion due to that force. The first law says that a body moving with a given speed will maintain that speed in a straight line unless a force acts upon it – no force means no acceleration.

The second law states that a force (F) will make a body accelerate by an amount (a) that is inversely proportional to its mass (m): $F = ma$. The third law says that whenever one body applies a force (the 'action' force) to a second one, the second one simultaneously applies an equal and opposite 'reaction' force on the first. Stepping off a boat onto a pier will make the boat move away, for instance.

Inspired by watching an apple falling from a tree, Newton reasoned from his laws of motion that there must be an attractive force between the Earth and the apple, which he called 'gravity'. This force could have huge range and be responsible for the orbit of the Moon around the Earth, if the Moon had just the right speed to remain in orbit despite constantly 'falling' towards the Earth.

He went on to show that the gravitational force between two massive objects is directly proportional to the product of their masses and weakens with the square of the distance between them. But troublingly, the theory didn't explain why the force was transmitted across empty space. This problem is resolved in **Einstein**'s general relativity theory, which views gravity as a natural consequence of the curvature of space–time.

Newton also showed that a prism could split white light into a spectrum of colours. He reasoned that this distortion should degrade the quality of refracting telescopes (those using a glass lens) and this inspired him to build the first practical reflecting telescope, using a mirror instead of a lens. He also shares credit with **Leibniz** for developing calculus, the branch of maths that deals with rates of change of quantities.

For all his scientific reason, Newton was a deeply religious man and actually wrote more on religion than he did on natural science. He viewed his scientific work as teasing out the plans of an orderly, divine creator: 'Gravity explains the motions of the planets, but it cannot explain who set the planets in motion. God governs all things and knows all that is or can be done.'

Newton famously had a complex personality, responding with prickliness to criticism and frequently falling out with rivals. In irritation when no one turned up to one of his lectures, he gave the lecture anyway to an empty room. He possibly was also capable of modesty: in a letter to his adversary **Hooke** in 1676, he complimented progress by both Hooke and Descartes, adding: 'If I have seen a little further it is by standing on the shoulders of giants'. More cynical observers have interpreted this as sarcasm.

In his fifties, Newton was appointed Warden then Master of the Royal Mint, and he took bold steps to prosecute counterfeiters. He was knighted in 1705. On his death in 1727, he was buried in Westminster Abbey.

EDMOND HALLEY
8 NOVEMBER 1656 – 14 JANUARY 1742

English astronomer, geophysicist and explorer best known for predicting the return of Halley's Comet

'In the year 1456, in the summer time, a comet was seen passing retrograde between the Earth and the Sun ... I dare venture to foretell that it will return again in the year 1758.'

Halley predicts the return of the famous comet now named after him
(from *Miscellanea curiosa*, 1706)

Halley was born in Shoreditch and developed an early interest in maths. From 1673, he studied at the University of Oxford, and he wrote papers on the solar system and sunspots while still an undergraduate. In 1675, he began work with the Astronomer Royal **Flamsteed**, assisting him with observations in Oxford and Greenwich. In 1696, Halley was appointed deputy controller of the Mint at Chester, largely through the influence of **Newton**. In 1704, he became Savilian Professor of Geometry at Oxford and, in 1720, he succeeded Flamsteed as Astronomer Royal, a position he held until his death.

Halley's first major project was to create a catalogue of southern stars from the island of St Helena in the South Atlantic. He left for St Helena in 1676 and established an observatory on the island. Despite poor weather and frequent cloud cover, he catalogued 341 stars and discovered a star cluster in the constellation Centaurus. He also studied weather patterns there and charted the variation of atmospheric pressure with altitude. On his return in 1678, he became an Oxford graduate without taking the degree examinations, and in the same year, aged 22, he became one of the youngest Fellows of the Royal Society.

In 1684, Halley met Newton to discuss **Kepler**'s laws of planetary motion and Newton explained how he had developed his laws of universal gravitation. Halley encouraged him to write his famous work the *Principia* (1687), which Halley edited. He also paid for the book to be printed and did much to promote Newton's work.

From around 1695, Halley studied comet orbits, which he suspected might be elliptical so that comets could make regular returns to the inner solar system. He calculated that a bright comet seen in 1682 was probably the same one that appeared in 1531 and 1607, and possibly also 1305, 1380 and 1456. He predicted that it would return again in December 1758. (Now known as Halley's Comet, it was sighted again on Christmas day 1758, although Halley did not live to see it.)

Halley's research was wide-ranging and adventurous. In 1698, he was given command of a small naval vessel to study geomagnetism in the South Atlantic and map the positions of coasts, ports and islands. He discovered the proper motion of stars, the way they move slightly sideways, viewed from the Earth. Until then, astronomers were not aware that stars moved relative to each other at all.

Halley's studies of the tides and the geomagnetic field set standards that lasted for centuries. He studied the magnetic origins of the northern lights (the *aurora borealis*) as well as the design and construction of diving bells. He pioneered the creation of quantitatively accurate mortality tables. He also proposed that by observing a transit of Venus, when Venus passes across the face of the Sun, it would be possible to work out the distance to the Sun accurately. Captain James Cook and others successfully did this in 1769, after Halley's death.

He had a famously stormy relationship with Flamsteed, who objected to Halley's appointment as professor at Oxford, complaining that Halley 'now talks, swears and drinks brandy like a sea captain'. Flamsteed was enraged at being put under pressure by Halley and Newton to publish the star catalogues he had compiled at Greenwich before they were complete. Halley forced him to hand over the incomplete observations and saw that they were edited and printed, after adding a preface that accused Flamsteed of lacking public spirit. Flamsteed later bought up 300 copies of the catalogue and burned them.

But Halley will remain best known for the comet that bears his name; most people get, at most, one chance to see it in their lifetime.

CARL LINNAEUS
(ORIGINALLY CARL NILSSON LINNÆUS IN SWEDISH)
23 May 1707 – 10 January 1778

Swedish naturalist and physician who laid the foundations for modern taxonomy

'The first step in wisdom is to know the things themselves; this notion consists in having a true idea of the objects; objects are distinguished and known by classifying them methodically and giving them appropriate names. Therefore, classification and name-giving will be the foundation of our science.'

From Linnaeus's *Systema Naturae* (1735)

Born in Råshult, Linnaeus studied medicine in Lund and botany in Uppsala, where he remained as a lecturer from 1730. From 1738, he practised as a doctor in Stockholm and in 1741 he became professor of medicine and botany at Uppsala University. Between 1732 and 1735, he travelled in Lapland and mainland Europe to meticulously document their flora and animal life.

To classify living organisms, Linnaeus introduced the binomial nomenclature system, in which each species has a generic name (the group to which it belongs) and then a specific species name. For instance, he coined the name *Homo sapiens* for modern humans. In his *Systema Naturae* (1758), he also detailed a basic classification system for animals, plants and microbes, which groups species according to shared physical characteristics in a branching tree-like hierarchy. This would later help establish the theory of evolution, something that Linnaeus ironically rejected.

In the taxonomy of Linnaeus there were three kingdoms (animal, vegetable or mineral) divided into more than 30 classes. The classes subdivided into orders, families, genera and then species, with an additional rank lower than species. The groupings have been revised since then to take account of new information about evolutionary trees. By the time Linnaeus died, he was one of the most revered scientists in the world.

DANIEL BERNOULLI
8 February 1700 – 8 March 1782

Dutch-born Swiss mathematician who founded the science of fluid dynamics

'There is no philosophy which is not founded upon knowledge of the phenomena, but to get any profit from this knowledge it is absolutely necessary to be a mathematician'

Quote attributed to Bernoulli

Bernoulli was born in Groningen, the Netherlands, to a Swiss family of mathematicians. He had a bitter and quarrelsome relationship with his overbearing father Johann Bernoulli, one of the early developers of calculus. Daniel grew up in Basel, and studied medicine and mathematics there as well as in Heidelberg and Strasbourg. He went to St Petersburg in 1724 as professor of mathematics, then in 1733 he returned to the University of Basel, where he held professorships in anatomy and botany, then physics.

His most famous work was *Hydrodynamica* (1738), a landmark text describing fluid flow and the fundamental relationships between pressure, density and velocity in the fluid. He also proposed what's now called Bernoulli's principle – the faster a fluid flows, the lower its pressure, an important concept in aviation. The curved upper surfaces of aircraft wings are shaped to force air to follow a longer path over the top of the wing, speeding it up. This lowers pressure above the wing and creates a net upward force, or lift.

Bernoulli's work was amazingly wide-ranging. He worked on theories of tides and magnetism, as well as methods for telling the time at sea, ocean currents and ways to stabilize ships during storms. He was also a close friend of **Euler**, who he worked with on problems of elasticity, and he demonstrated that **Boyle**'s law follows naturally if gases consist of small particles bouncing elastically off each other in ceaseless motion.

LEONHARD EULER
15 April 1707 – 18 September 1783

Pioneering and prolific Swiss mathematician and physicist

'Euler calculated without apparent effort, as men breathe,
or as eagles sustain themselves in the wind'

Euler described by French astronomer Dominique Arago

Born in Basel, Euler had a gift for maths from a young age and enrolled at the University of Basel aged just 13. In 1723, he received his MPhil degree for a thesis comparing the theories of **Newton** and **Descartes**. He then worked with his friend **Bernoulli** in St Petersburg, Russia, and in 1741 he took a post at Frederick the Great of Prussia's Berlin Academy. He returned to Russia in 1766, invited by Catherine the Great.

Euler made important discoveries in just about all areas of mathematics, including graph theory and calculus, which he developed by integrating Newton and **Leibniz**'s methods to make calculus easier to apply to real-world problems. Many notations that he introduced are still used today, including f(x) – meaning a function f applied to a variable x – and the letter i to denote the imaginary unit (the square root of negative one, which is non-existent but useful in mathematical calculations).

Over his lifetime, Euler published around 800 papers. In physics, his work helped promote the wave theory of light proposed by **Huygens**; until the 1740s, the prevailing view was Newton's notion that light was 'corpuscular', made up of little particles. (The modern view from quantum mechanics is that light has both these properties, the upshot of 'wave–particle duality'.) Euler also recognized that light's colour depends on its wavelength, and in astronomy, he improved methods for analyzing orbits of bodies such as the Moon.

JOSEPH BLACK
16 April 1728 – 6 December 1799

Scottish physician famous for discovering carbon dioxide and the concept of latent heat

'When ice, for example, or any other solid substance, is changing into a fluid by heat, I am of opinion that it receives a much greater quantity of heat than that what is perceptible in it immediately after by the thermometer'

Black describes his discovery of latent heat in
Lectures on the Elements of Chemistry (1803)

Black was born to a Scottish mother in Bordeaux, France, where his Irish father was a wine merchant. Aged 12, he was sent to school in Belfast to learn Latin and Greek. He enrolled at Glasgow University in 1744 to study the arts, but later switched to medicine. He also learned chemistry, which had just begun to be taught at Glasgow University. Black moved to Edinburgh in 1752 to further his medical studies, and from 1756 he held professorships at Glasgow and Edinburgh universities.

He is best known for discoveries in chemistry, including the concept of latent heat, the heat absorbed or released during a phase transition. He showed in 1761 that heating ice at its melting point does not raise the temperature of the ice–water mixture, but simply increases the proportion of water. He also showed that different chemicals have different specific heats, the amount of heat energy per unit mass needed to raise the temperature by 1°C (1.8°F). However, the reason for this was unclear until scientists including **Maxwell** developed a kinetic theory of heat.

Black also discovered carbon dioxide. When he heated the compound we now call calcium carbonate or poured acid on it, he noticed it released a gas (he called it 'fixed air') in which candles wouldn't burn. He also worked on steam experiments with James Watt, who developed the steam engine.

GEORGES-LOUIS LECLERC, COMTE DE BUFFON
7 September 1707 – 16 April 1788

French naturalist, mathematician and prolific writer who developed early ideas about evolution

'Only those works which are well-written will pass to posterity:
the amount of knowledge, the uniqueness of the facts, even
the novelty of the discoveries are no guarantees of immortality ...
These things are exterior to a man but style is the man himself.'

From Buffon's *Natural History* (vol 7, 1753)

Buffon was born into a wealthy family in Montbard, in France's Bourgogne region. He attended the universities of Dijon and Angers, first studying law but developing a strong interest in maths and astronomy. In 1739 he was appointed keeper of the Jardin du Roi, the botanical gardens in Paris, and from then he had a lifelong interest in natural history.

Buffon is most famous for compiling vast amounts of information about the entire natural world. From 1752, he and various collaborators published around 40 volumes of his *Natural History* series, part of which appeared after his death. It was clearly written for the general public, translated into many languages and made him one of the most widely read authors in the world. He also translated **Newton**'s work on calculus into French.

He speculated that the Earth may have existed for 75,000 years – a bold suggestion at the time in Christian Europe, where the Biblical age of 6,000 years was still unquestioningly accepted. Buffon also developed a rudimentary theory of evolution, recognizing that different regions of the world have distinctive animals and plants, even if their climate is similar. He noted that some animals had features that were useless to them, and suggested that species may have 'improved' or 'degenerated' after migrating out from a common site.

BENJAMIN FRANKLIN
17 January 1706 – 17 April 1790

*American political leader, scientist and polymath who
pioneered early studies of electricity*

'The electrical matter consists of particles extremely subtile, since it
can permeate common matter, even the densest metals, with such
ease and freedom as not to receive any perceptible resistance. If
anyone should doubt whether the electrical matter passes through
the substance of bodies, or only over along their surfaces, a shock
from an electrified large glass jar, taken through his own body, will
probably convince him.'

Franklin quoted in *Benjamin Franklin's Experiments* (1941)

Born in Boston, Franklin was apprenticed to
his brother as an assistant printer at the age of 12.
He later established his own printing business and
became a successful newspaper editor. He moved
into public affairs in 1936 as Clerk of the State
Assembly and later Deputy Postmaster-General
for the colonies. He went on to become one of the
US's most influential Founding Fathers and the only
statesman to have signed the four documents that created the nation,
including the Declaration of Independence.

Despite his busy political life, Franklin made important contributions
to science. He did early experiments on electricity, which he perceived to be
a fluid, and introduced the terms 'positive' and 'negative' to describe its two
possible states. Suspecting that lightning is an electrical phenomenon, he
conducted his famous kite experiment in 1752, according to a later account
by **Priestley**. He flew a kite with a wet string during a thunderstorm and
showed it conducted electricity with visible sparks. After that he proposed
the use of tall lightning rods on buildings, wired to the ground, to protect
them from lightning damage.

The diverse range of other problems Franklin studied included light,
heat and the dynamics of ocean currents and their effects on weather. His
many inventions included an efficient stove and bifocal glasses.

HENRY CAVENDISH
10 October 1731 – 24 February 1810

British scientist who discovered hydrogen and made the first laboratory measurement of the force of gravity

'[Cavendish] fixed the weight of the Earth; he established the proportions of the constituents of the air; he occupied himself with the quantitative study of the laws of heat; and lastly, he demonstrated the nature of water and determined its volumetric composition. Earth, air, fire, and water – each and all came within the range of his observations.'

From *Essays in Historical Chemistry* by Thomas Edward Thorpe (1894)

Born into a wealthy English family in Nice, France, Cavendish studied for four years at Cambridge University from 1749 but left without graduating. Historians suggest this is partly because he objected to the obligatory religious exercises involved at the time, and partly because he had a debilitating shyness that made him unable to face his professors during his examinations.

With a comfortable allowance and later an inheritance from his family, Cavendish then devoted almost 60 years to first-class scientific research, the exclusive passion of his life. He showed no particular desire for publishing his work or seeking credit for it, seemingly motivated solely to satisfy his own curiosity.

By reacting metals with acids, Cavendish isolated and studied hydrogen and showed that it reacts with oxygen to form water. Although other chemists had done this earlier, Cavendish was first to identify this 'inflammable air' as an element, which was later named hydrogen by **Lavoisier**. Cavendish reported his early research on this to the Royal Society in 1766. He was also first to determine the ratio of oxygen (then called 'dephlogisticated air') to nitrogen ('phlogisticated air') in the atmosphere, correctly concluding that it is about 1:4.

He demonstrated that once he removed oxygen and nitrogen from air, there was a tiny amount of gas left over that was no more than 1/120th of the total. He recognized that it must contain a gas that is highly resistant to reaction. Later **Ramsay** confirmed this with his discovery of the inert gas argon, which is the third most abundant constituent of the atmosphere, albeit in tiny amounts.

He is also remembered for the 'Cavendish experiment', an experiment devised by English geologist John Michell who died before he could carry it out. In this work, Cavendish measured the gravitational force between two masses in a laboratory for the first time. In 1798, he suspended a lead ball from each end of a light rod and by bringing very large balls near the two light balls, he could measure the slight twist of the rod due to the force of gravity between the large and small balls, from which he could calculate the gravitational force between them.

From his data, Cavendish used **Newton**'s equations to calculate the density of the Earth. He concluded that the Earth's average density as 5.48 times that of water (the accepted value today is 5.518), although he fluffed the calculation – his data actually point to a smaller and less accurate value of 5.45. Later, Cavendish's measurements were used to calculate a fairly accurate value of G, the gravitational constant that appears in Newton's equations.

Cavendish also did extensive research on electricity, discovering the relationship between voltage and current (now called Ohm's law) and the inverse square law of electric force with distance (Coulomb's law). However, much of this work wasn't acknowledged until **Maxwell** rediscovered Cavendish's unpublished work, by which time other scientists were credited with the discoveries.

ALBRECHT VON HALLER
16 October 1708 – 12 December 1777

Swiss naturalist and anatomist often considered the father of experimental neurology

'I call that part of the human body irritable, which becomes shorter upon being touched; very irritable if it contracts upon a slight touch, and the contrary if by a violent touch it contracts but little. I call that a sensible part of the human body, which upon being touched transmits the impression of it to the soul.'

From Haller's *A Treatise on the Sensible and Irritable Parts of Animals* (1936)

Haller was born in Bern and developed a precocious intellect as a child. From the age of 16, he studied at the universities of Tübingen and Leiden, where he graduated in 1727. In 1736, he was appointed professor of anatomy, botany and medicine at the University of Göttingen where he remained until 1753.

Haller was a prolific poet and wrote several works on politics as well as an extensive work on the flora of Switzerland, but his most important contributions to science involved experimental work on muscles and nerves, in which he distinguished between the previously muddled realms of movement and sensation. He showed that a stimulus to a nerve can trigger the sharp contraction of a muscle, and that muscles do not experience a sensation themselves but carry the impulses that transmit sensation to the spinal cord or brain.

He experimented on animals by stimulating or damaging various parts of the brain, then measuring the response or paralysis that resulted, for which he is regarded as a founder of modern neurology.

LUIGI ALOISIO GALVANI
9 September 1737 – 4 December 1798

Italian anatomist who made early studies of electrical signals from the nervous system

'While one of those who were assisting me touched lightly, and by chance, the point of his scalpel to the internal crural nerves of the frog, suddenly all the muscles of its limbs were seen to be so contracted that they seemed to have fallen into tonic convulsions'

Quoted in *Makers of Electricity* by Michael Francis O'Reilly (1909)

Born in Bologna, Galvani studied theology but later followed his father into medicine after studying at Bologna's medical school. He became an anatomy lecturer at the University of Bologna in 1762 and was appointed professor there in 1775.

Around 1771, Galvani noticed that legs from dead frogs twitched when an electrical current was applied to them, demonstrating that electrical energy provides the impetus behind muscle movement. Reasoning that lightning is an electrical phenomenon, as **Franklin** had suggested, Galvani hung frog legs on metal hooks outside a window during a thunderstorm predicting that the lightning would make them twitch. Curiously, however, the legs still twitched even when the storm had passed, providing they were in contact with two different metals.

Galvani incorrectly concluded that the activating force came from the frog legs themselves rather than the metals, and he named it 'animal electricity'. The Italian physicist Alessandro Volta later proved that he was wrong; the electric current arose from the metals (because different metals differ in their tendency to lose electrons). Galvani's name nonetheless became immortalized in related electrical effects such as the Galvanic cell and galvanization, the process of applying a protective zinc coating to a metal like iron to prevent corrosion. A crater on the Moon is also named after him.

CHARLES MESSIER
26 June 1730 – 12 April 1817

French astronomer famous for his catalogue of fuzzy objects in the night sky

'What caused me to undertake the catalogue was the nebula I discovered above the southern horn of Taurus on September 12, 1758, while observing the comet of that year. This nebula had such a resemblance to a comet in its form and brightness that I endeavoured to find others so that astronomers would not confuse these same nebulae with comets just appearing.'

From Messier's journal *Connaissance des Temps* (1801); he describes what's known today as the Crab Nebula

Born in Badonviller, Lorraine, Messier developed a childhood interest in astronomy and in 1751 became an assistant to Joseph Nicolas Delisle, astronomer to the French Navy. Messier was the first astronomer in France to see the famous 1758 return of **Halley**'s Comet (Halley had predicted the comet's return but did not live to see it).

Inspired by this observation, Messier devoted his life to hunting for comets and he discovered more than a dozen. However, he is most famous for compiling a catalogue of objects that were not comets, which move fast relative to background stars. He identified stationary fuzzy patches, which lie at a great distance, and catalogued them so that he and other astronomers could rule them out in comet searches.

Professional and amateur astronomers still refer to the astronomical objects in Messier's catalogue, from M1 to M110, by their Messier numbers. Messier had no clue as to what these objects actually were, however. But because he selected objects that are distant, yet nonetheless extended and bright enough for him to see with relatively primitive telescopes, his catalogue was biased to include the most spectacular objects visible from the Earth in modern times, including galaxies, star clusters and the remnants of exploded stars.

JOSEPH LOUIS LAGRANGE
25 JANUARY 1736 – 10 APRIL 1813

*French–Italian mathematician and astronomer who
solved complex gravity problems*

'Joseph Louis Lagrange was the greatest mathematician of the 18th
century ... In appearance he was of medium height, and slightly formed,
with pale blue eyes and a colourless complexion. In character he was
nervous and timid, he detested controversy, and to avoid it willingly
allowed others to take the credit for what he himself had done.'

From *A Short Account of the History of Mathematics* by W.W. Rouse Ball (1908)

Lagrange was born Giuseppe Lodovico
(Luigi) Lagrangia in Turin, having both French and
Italian descent. Inspired by reading a paper by **Halley**,
he educated himself in maths and was appointed as
a lecturer in Turin's Artillery School. In 1766, he
succeeded **Euler** as director of mathematics at the
Berlin Academy, where he remained for more than
20 years. In 1787, he moved to Paris at the invitation
of Louis XVI, and under Napoleon he became a senator and a count.

Early in his career, Lagrange invented a generalized form of calculus,
the calculus of variations, which later became important for solving
physics problems in a wide range of fields including gravitation and
quantum mechanics. On several occasions, he won the Grand Prize from
the French Royal Academy of Sciences for solving contemporary problems
in physics, such as explaining the gravitational forces that make just one
side of the Moon permanently face the Earth, and as well as solving more
complex gravitational problems involving several bodies, such as Jupiter,
its four satellites known at the time, and the Sun. He also identified the
five 'Lagrangian points' in an orbital configuration where a small object
such as a spacecraft can remain stationary relative to two larger objects,
such as the Earth and Moon.

His major work *Mecanique Analytique* (1788) applied calculus to the
mechanics of rigid bodies and developed elegant general equations to describe
motion in mechanical systems. Lagrange lost interest in maths during a bout
of depression in Paris, but regained it after the French Revolution.

CARL WILHELM SCHEELE
9 December 1742 – 21 May 1786

German–Swedish chemist who discovered many elements including oxygen and hydrogen

'I filled a ten-ounce glass with this air and put a small burning candle into it; when immediately the candle burnt with a large flame, of so vivid a light that it dazzled the eyes. I mixed one part of this air with three parts of air, wherein fire would not burn; and this mixture afforded air, in every respect familiar to the common sort.'

From Scheele's *Chemical Observations and Experiments on Air and Fire* (1777)

Scheele was born in Stralsund, Germany, a region that was then under Swedish rule. He was given little formal education but as a young teenager he became an apprentice to an apothecary in Gothenburg, where he was free to experiment with chemicals and read up on scientific texts. In 1768, he moved to Stockholm and again worked in a pharmacy before becoming a laboratory assistant in Uppsala from 1770.

In Uppsala, Scheele discovered several reactions that released 'fire air' (oxygen) in the early 1770s. One way was to heat silver carbonate to release carbon dioxide and oxygen, and then use an alkali to mop up the carbon dioxide component. He recognized that while the atmosphere contains oxygen, which encourages burning, it also includes another component ('foul air', or nitrogen) that prevents it. Scheele isolated oxygen around two years before **Priestley**; however, it was Priestley who published his finding first.

In 1775, Scheele was elected as a member of the Swedish Royal Academy of Sciences and in the same year he moved to Köping, where he was appointed as superintendent of the pharmacy. During his lifetime he discovered a vast range of elements and compounds, including chlorine, hydrogen sulphide and hydrogen cyanide. He isolated and studied uric acid, glycerol, lactic acid and copper arsenite, and he also noted that light can alter silver salts (converting them to metallic silver) – this reaction became the mechanism behind film photography.

Scheele died at the age of 43. His poor health was probably related to his exposure to laboratory toxins, including poisonous cyanide compounds.

JOSEPH PRIESTLEY
13 March 1733 – 6 February 1804

English chemist, natural philosopher and theologian who discovered oxygen

'I procured a mouse, and put it into a glass vessel, containing two ounce-measures of the air from mercurius calcinatus. Had it been common air, a full-grown mouse, as this was, would have lived in it about a quarter of an hour. In this air, however, my mouse lived a full half hour; and though it was taken out seemingly dead, it appeared to have been only exceedingly chilled; for, upon being held to the fire, it presently revived.'

Priestley describes the effects of oxygen-rich air

Priestley was born in Birstall in Yorkshire. He studied theology, enrolling at the Dissenting academy in Daventry in 1752. As both a minister and teacher, he lived at various locations in England but in 1791, his house was burned by a mob, enraged at his radical theological views and support for the French Revolution. He later moved to the US and lived in Pennsylvania for the last decade of his life.

Priestley wrote on a wide range of topics, including politics, but is best remembered for his work in experimental science. He showed that if a candle burned in air until extinguished, a sprig of mint could 'purify' the air so the candle could be relit. He realized that the mint released the purifying component and had he pursued this line of enquiry he might have discovered photosynthesis, but that was left to **Ingenhousz** and others. In work on electricity, Priestley showed that different materials display a continuum of values of electrical conductivity, from metals to pure insulators.

He also isolated the gas oxygen in 1774. **Scheele** had done this two years earlier, but Priestley was first to publish the finding. When he heated 'mercurius calcinatus' (mercury oxide), he showed it liberated a colourless gas in which a candle burned extremely brightly. On a visit to Paris, he discussed this with **Lavoisier**, who later repeated and improved the experiment, and correctly understood the discovery's full significance.

ANTOINE-LAURENT DE LAVOISIER
26 August 1743 – 8 May 1794

French scientist often considered the father of rigorous modern chemistry

'Chemists have made of phlogiston a vague principle that has not been rigorously defined and, hence, it can be adapted to any explanation one wants to give for it. Sometimes the principle is heavy, sometimes not; sometimes it is free fire, sometimes it is fire combined with the earthy element ... It is a veritable Proteus that changes form every second. Chemistry must be subjected to more rigorous reasoning, the facts that daily enrich this science must be stripped away from what mere arguments and prejudices add to it; what comes from fact and observation must be separated from what is systematic and hypothetical.'

Lavoisier quoted in *Lavoisier: Chemist, Biologist, Economist* by Jean-Pierre Poirier and Rebecca Balinski (1998)

Lavoisier came from a wealthy family in Paris. In his family's tradition, he studied law, but he was much more interested in science and learned from France's most eminent scientists in a wide range of the fields from maths to geology, astronomy and chemistry. He successfully integrated himself into leading French intellectual circles and was elected to the French Royal Academy of Sciences in 1768, when he was aged only 25.

By this time, he had already identified the composition of gypsum (plaster of Paris), showing by meticulous measurements that it contained water, created a geological map of France and studied the feasibility of installing street lamps in French towns. To fund his studies, he invested in and worked for a private tax-collection company. This decision would later turn out to have fatal consequences.

In 1768, Lavoisier dispelled the ancient notion that earth could be created from pure water (this misconception had arisen because a solid sediment of impurities can appear when water evaporates from a container). In careful measurements of a container of water's weight before and after evaporation, Lavoisier proved that the sediment had come from the container, not the water.

In all his work, Lavoisier had a strong conviction about the conservation of matter, that it can be converted, but never created or destroyed. He also played a key role in efforts to standardize weights and measures, helping establish the metric system of units.

But Lavoisier is probably most famous for his work on gases, in which he finally understood the basic mechanisms of compound formation. Until his time, it was commonly understood that when substances burn in chemical combustion, a chemical called 'phlogiston' is released. In discussions with **Priestley**, he learned that Priestley had isolated a gas that made a candle burn amazingly brightly. The English scientist had named it 'dephlogisticated air', assuming its properties reflected the fact that it absorbed phlogiston very readily.

Lavoisier put two and two together. His precise mass measurements revealed that combusting materials do not actually lose mass, they combine with a gas in the air that actually increases their weight. He renamed Priestley's gas 'oxygen' and argued that air also contains a second gas that does not support combustion, calling it 'azote' (today known as 'nitrogen').

From the mid-1770s, Lavoisier was in charge of the government's gunpowder manufacturing at the Royal Arsenal, where he set up an advanced chemical laboratory. Working with the physicist **Laplace**, he investigated animal respiration and showed that body temperature depends on oxygen intake and that nitrogen plays no part in supporting animal life. In 1787, he introduced the modern symbol of chemical notation, indicating that compound names should express their ingredients (for instance, carbon dioxide contains carbon and oxygen). His *Elementary Treatise on Chemistry* (1789) is often seen as the first modern chemistry textbook, establishing **Boyle**'s definition of an element as the most basic chemical unit. He also repeated experiments by **Cavendish** that had produced 'inflammable air'; Lavoisier named it hydrogen.

Following the outbreak of the French Revolution in 1789, Lavoisier was targeted by revolutionaries due partly to his tax-collection activities and partly because he had condemned the scientific work of Jean-Paul Marat, a leading figure in the Terror. Lavoisier fled his home but was later arrested and charged with 'counter-revolutionary activity'. He was sentenced to death and guillotined on 8 May 1794. On his death, **Lagrange** commented: 'Only a moment to cut off that head and a hundred years may not give us another like it.'

(FREDERICK) WILLIAM HERSCHEL
15 November 1738 – 25 August 1822

German–British astronomer who discovered the planet Uranus

'It has generally been supposed that it was a lucky accident that brought this new star to my view; this is an evident mistake. In the regular manner I examined every star of the heavens ... it was that night its turn to be discovered. I had gradually perused the great Volume of the Author of Nature and was now come to the page which contained the seventh Planet.'

Herschel recalls his discovery of Uranus in a letter
(recounted in *Planets Beyond* by Mark Littmann, 2004)

Herschel was born in Hanover, Germany, where his father was a musician for the Hanoverian guard. He became a talented musician himself, and moved to England in 1757, where he became a well-known organist and music teacher in Bath from 1766.

Familiarity with the theory of sound led him to study maths and optics then to **Newton**'s theories, which gave him a burning desire to study the heavens with optical instruments he built with his sister Caroline, who moved to England from Germany in 1772.

Entirely self-taught in science, Herschel went on to become one of the most famous astronomers of his era. Through tireless grinding of lenses and polishing mirrors, the Herschels built the best reflecting telescope of its time. William Herschel wrote prolifically about his observations of many objects from the Moon to variable stars. He also investigated sunspots and speculated whether they influence climate and agriculture on Earth.

From 1781, he corresponded with the Royal Society about his observations of the rotation of planets and their moons. He noticed white spots near the poles of Mars, and correctly concluded that they were polar ice. Also in 1781, Herschel discovered Uranus. He noticed a patch of light in the sky that was too extended to be a point-like star. At first he assumed it was a comet, but his telescope observations revealed that it had a sharp edge rather than a fuzzy one typical for a comet. When enough observations had been gathered to compute the object's orbit, it turned out to be a circular orbit characteristic of a planet and that it must orbit the Sun beyond Saturn.

This was the first discovery of a new planet in history. All the other planets known at the time are visible to the naked eye and so had been known since ancient times. Although a few other astronomers had recorded Uranus in star maps, Herschel's observations were the first ones to recognize its planetary nature. Later, with even more powerful telescopes, Herschel discovered two moons of Uranus and two moons of Saturn.

In 1782, King George III invited Herschel to Windsor to become his private astronomer with Caroline as his assistant. From then on, they carried out their observations from Slough with even more vigour. William discovered around 800 double stars and showed that they were genuinely orbiting closely around each other; previously, astronomers had assumed they were coincidentally close to each other seen from Earth, but at vastly different distances. His measurements of double stars proved Newton's laws correct beyond the solar system for the first time.

He also proved that the Sun moves through the Milky Way, and therefore can't be considered to lie at its centre. Just as **Copernicus** showed that the Earth is not the centre of the Universe, Herschel proved that neither is the Sun. He was knighted in 1816.

JAN INGENHOUSZ
8 December 1730 – 7 September 1799

Dutch biologist who discovered photosynthesis in plants

'The production of dephlogisticated air [oxygen] from leaves is not
owing to the warmth of the Sun, but chiefly, if not only, to the light ...
No dephlogisticated air is obtained in a warm room, if the Sun does
not shine upon the jar containing the leaves.'

From Ingenhousz's *Experiments upon Vegetables* (1779)

Born in Breda, Ingenhousz studied medicine at the University of
Louvain, where he earned his MD degree in 1753. After further study in
Leiden and Edinburgh, he set up a medical practice in Breda before moving
to London, where he learned the new technique of smallpox inoculation.
He later successfully inoculated the Habsburg family in Vienna in 1768,
after which he was appointed personal physician to the Austrian Empress
Maria Theresa with an income that allowed him financial independence
for his scientific work.

Ingenhousz is best known for discovering photosynthesis in 1779.
Earlier, **Priestley** had noted that plants somehow made air breathable again
after it had been made toxic to animals by a burning candle. Investigating
this in hundreds of meticulous experiments on plants in water, Ingenhousz
showed that green plants give off a gas that he identified as oxygen (still
known at the time as 'dephlogisticated air'), while this process stops in the
dark, when they instead give off carbon dioxide.

His research also demonstrated that respiration occurs in all parts of
plants, including flowers and roots, and that plants produce much more
oxygen than they consume, so that animal and plant life mutually support
each other. Throughout his life, Ingenhousz travelled widely, meeting
and corresponding with **Benjamin Franklin** and **Cavendish**. His diverse
research also covered electricity, heat conduction and chemistry.

PIERRE SIMON DE LAPLACE
23 March 1749 – 5 March 1827

French mathematician and astronomer who analysed planetary orbits

'The weight of evidence for an extraordinary claim must be proportioned to its strangeness'

Quote often attributed to Laplace

Laplace was born in Beaumont-en-Auge, Normandy, and studied theology at Caen University from the age of 16. But instead of theology he developed a passion for mathematics and an impressive talent in the subject that led to him being appointed professor of mathematics at the Military School in Paris.

Building on work by **Lagrange**, Laplace showed that some puzzling anomalies in the motions of Jupiter and Saturn could be explained by their mutual gravitational pull. He also demonstrated that as long as the solar system remained isolated and the Sun didn't change its nature drastically, the orbits of the planets would remain stable for an indefinite period.

Favoured by Napoleon Bonaparte, a keen promoter of maths and science, Laplace was briefly appointed French minister of the interior. On reading Laplace's major work on astronomy, *Celestial Mechanics* (published in parts between 1799 and 1825), Napoleon is supposed to have remarked that he saw no mention of God in it, to which Laplace famously replied: 'I have no need of that hypothesis.'

Laplace also developed the nebular hypothesis of the solar system's formation, a theory still favoured today. It describes how the solar system evolved from a cloud of rotating gas that contracted to form the Sun with a swirling disc around it, in which the planets coalesced under gravity. In pure maths, he laid the foundations of modern probability theory and made key developments in differential equations.

Scottish scientist often considered the father of modern geology

'The past history of our globe must be explained by what can be seen
to be happening now. No powers are to be employed that are not
natural to the globe, no action to be admitted except those of which
we know the principle.'

From Hutton's paper 'Theory of the Earth'
(*Transactions of the Royal Society of Edinburgh*, 1785)

Born in Edinburgh, Hutton studied classics at
the University of Edinburgh and became a lawyer's
apprentice, but he had developed a strong interest
in sciences and switched to medicine. He attended
medical lectures in Edinburgh and in Paris before
earning his doctorate in Leiden in 1749 for a thesis
on blood circulation. After devoting several years to
agriculture and industry, he returned to Edinburgh
in 1768. Financially independent from business and property interests, he
was able to carry out his own scientific research, particularly in geology,
for the rest of his life.

Travelling throughout Britain to study rock formations, Hutton
developed his geological theories over many years without any sense of
pressure to publish them quickly. His best-known theory (later dubbed
'uniformitarianism') first appeared in print in a paper around 1788
and was extended into a two-volume book, *Theory of the Earth*, published
in 1795.

Hutton's key advance was to recognize that the geological processes
that shaped the Earth in the distant past are exactly the same as the ones
still occurring today. So current rates of processes such as erosion or
sedimentation made it possible to create a timeline of the past, indicating
how long it took to deposit sandstone of a given thickness, for example.
Geology could now become a true science.

A prevailing view before Hutton's work was that all rocks had been laid
down as mineral deposits in the oceans of the early Earth. Hutton realized

that granite rocks must have formed from the cooling of molten rock, and that the Earth's internal heat must gradually fuse sedimentary rocks to create rock types such as granite and flint. With the Earth's internal heat acting as the engine, a gradual and continuous rock cycle must account for the upheaval of rock strata, and their folding and tilting, with the uplift of mountains.

Hutton envisaged countless cycles of sediment deposition on seabeds followed by uplift and erosion. As he put it: 'The result, therefore, of our present enquiry is, that we find no vestige of a beginning, no prospect of an end.' His theory required a much larger age for the Earth than traditional Biblical interpretations implied. Rather than being a few thousand years old, it must be millions of years old. Although this was still a vast underestimate compared to modern values (the Earth is now thought to be about 4.5 billion years old), this nonetheless marked a key milestone in rational geological debate.

Hutton also developed a theory of the Earth's water cycle, arguing that rainfall is regulated by air humidity as well as the meeting of various air currents high in the atmosphere. Extending his uniformitarian concept of gradual change to animals and plants, he also developed an early theory of evolution, suggesting that life that is best adapted to its environment would thrive and reproduce, while animals and plants that are least well adapted would die out. While Hutton thought this process could stimulate subtle change, he rejected the notion that evolution could produce radically different species.

He was renowned for his cheerful and amiable character. Hutton was a close friend of **Black**, and together with the famous economist Adam Smith they founded the 'Oyster Club' in Edinburgh for lively weekly meetings of local and visiting intellectuals. But he had a peculiar and obscure writing style that probably delayed acceptance of his theories and gave ammunition to his critics, and this meant he possibly received less acclaim in his lifetime than he deserved. The Scottish scientist John Playfair communicated Hutton's theories more successfully and **Lyell** later extended and adapted them, leading to their wide acceptance. Lyell's work, which **Darwin** read during his voyage on the Beagle, was an important influence on Darwin when he developed his theory of evolution.

EDWARD ANTHONY JENNER
17 MAY 1749 – 26 JANUARY 1823

English scientist who developed the smallpox vaccine

'While the vaccine discovery was progressive, the joy I felt at the prospect of being the instrument destined to cast away from the world one of its greatest calamities [smallpox], blended with the fond hope of enjoying independence and domestic peace and happiness, was often so excessive, that, in pursuing my favourite subject among the meadows, I have sometimes found myself in a kind of reverie. It is pleasant to me to recollect, that these reflections always ended in devout acknowledgement to that Being, from whom this and all other mercies flow.'

Jenner quoted in *The Life of Edward Jenner* by John Baron (1827)

Born in Berkeley, Gloucestershire, Jenner became an apprentice to a surgeon from the age of 13. From 1770 to 1772, he trained in medicine at St George's Hospital in London under John Hunter, one of the most famous surgeons in England and a well-respected experimental scientist. Jenner then returned to his home county where he worked as a general practitioner and surgeon, and also developed his interests in natural history. In 1792, he went on to earn his MD degree from St Andrews University.

Jenner's research was extremely wide ranging. He studied geology, carried out experiments on human blood and designed and built a hydrogen balloon. In bird studies, he showed that cuckoo hatchlings, rather than adult cuckoos, are responsible for throwing eggs and chicks of their foster parents out of the nest.

But Jenner is best known for his work on smallpox, a devastating disease thought to have emerged as far back as 10,000 BC. During epidemics, mortality rates reached about 60 per cent, and the infant mortality rate from smallpox was even higher. Records suggest that during the early decline of the Roman Empire, a single epidemic of the disease killed almost 7 million people.

By Jenner's time, there was already a widespread practice of protecting people from the disease by 'variolation' – deliberately infecting the skin with smallpox using pus from a patient in a controlled manner to induce a mild form of the disease that would confer future immunity. However, this ran the risk of inducing full-blown smallpox and also the risk of people unwittingly transmitting the disease to others. An estimated 2 to 3 per cent of variolated people died from the disease, became the source of another epidemic, or suffered from other infectious diseases such as tuberculosis transmitted by accident during the procedure.

Jenner heard of anecdotal reports that milkmaids infected with cowpox never contracted smallpox later. In people, cowpox causes red blisters but the disease is mild. In May 1796, Jenner took swabs of fresh cowpox blister from a dairymaid and used them to inoculate an 8-year-old boy. He developed a mild illness but recovered, and when Jenner inoculated him again with pus from a fresh smallpox lesion, he developed no disease at all, suggesting he was completely protected.

Jenner decided to call the procedure vaccination (from the Latin *vaccinus*, meaning of or from cows). Moreover, he demonstrated that the protective cowpox could be effectively inoculated from person to person, not just directly from cattle. After repeating this experiment on more than 20 volunteers, Jenner's vaccination technique eventually became accepted. He was widely ridiculed. Critics, especially religious ones, viewed it as ungodly and repulsive to vaccinate someone with material from sick animals. A satirical cartoon from 1802 depicted vaccinated people sprouting cow's heads.

But vaccination became common as its success became obvious. The British government banned variolation in 1940, thereafter providing cowpox vaccination free of charge. In 1979, the World Health Organization declared smallpox an eradicated disease.

Jenner had no explanation for why his method worked – no one could see the virus with the microscopes of his era. He was also probably not even the first to use the vaccination concept against smallpox. There's evidence that at least a few individuals learned to protect people in rural dairy communities from smallpox using cowpox. But it was Jenner's relentless promotion and persistent research that led to safe mass immunization programmes. His vaccine also laid the groundwork for many modern-day discoveries in immunology. Jenner is also credited with making important steps in understanding the cause of angina.

JOHANN CARL FRIEDRICH GAUSS
30 April 1777 – 23 February 1855

German mathematician and natural philosopher regarded as one of the greatest mathematicians of all time

'It is not knowledge, but the act of learning, not possession but the act of getting there, which grants the greatest enjoyment. When I have clarified and exhausted a subject, then I turn away from it, in order to go into darkness again. The never-satisfied man is so strange; if he has completed a structure, then it is not in order to dwell in it peacefully, but in order to begin another. I imagine the world conqueror must feel thus, who, after one kingdom is scarcely conquered, stretches out his arms for others.'

From a letter from Gauss to Hungarian mathematician Farkas Bolyai (1808)

Born into a peasant family in Brunswick, Gauss developed an astonishing talent in arithmetic even by the age of three, and his mathematical genius later led the Duke of Brunswick to arrange his education at the Caroline College in Brunswick and then at the University of Göttingen from 1795. In 1806, he was appointed director of the observatory at Göttingen, where he remained for the rest of his life.

In 1801, Gauss published *Disquisitiones Arithmeticae*, which is often regarded as heralding the start of modern number theory. This fused his own work with that of his predecessors. It elegantly described many advances, including Gauss's solution to a famous problem in Greek geometry – the inscription of regular polygons in a circle. He showed that a polygon with 17 sides can be accurately imposed on a circle using a ruler and compasses, and then generalized this result to show that this can always be done provided the number of polygon sides is a prime number with the form $2^{2m} + 1$, where m is a positive integer.

Among many other mathematical achievements, Gauss proved the fundamental theorem of algebra. This states that every equation of degree n with complex coefficients has at least one root that is itself a complex number (an imaginary number that's a multiple of the square root of -1).

In astronomy, Gauss determined the orbit of the huge asteroid Ceres (now redefined as a dwarf planet) using observations by the Italian astronomer Giuseppe Piazzi, who had discovered the body on 1 January 1801 then 'lost' it. Ceres was rediscovered after Gauss successfully predicted its position. His success in doing this encouraged him to do further work on orbit determination which he published in his 1809 work, *Theoria Motus Corporum Coelestium*. This book elegantly streamlined previously cumbersome maths of the 18th century for orbital prediction.

To calculate planetary orbits, Gauss used the method of least squares, an accurate means of exploiting all available observations to find a solution with the smallest possible error. His awareness of the importance of minimizing errors led him to derive the Gaussian law of error, familiar to statistics students as the 'normal distribution' bell curve.

He was also among the first people to study the geometry of curved spaces. Previously, mathematicians had only considered Euclidean geometry (flat spaces, in which parallel lines never meet; see **Euclid**). However, Gauss left little writing on this subject so it's unclear how far he developed it. But non-Euclidean geometry later became a cornerstone of much 20th-century research, including **Einstein**'s theory of general relativity.

One puzzle he solved was particularly personal – the date of his own birth. Gauss's illiterate mother never recorded his birth date but recalled that he had been born on a Wednesday, eight days before the Feast of the Ascension, which occurs 40 days after Easter. Gauss pinpointed his own birth date by deriving a formula to calculate the date on which Easter falls in both past and future years. This pinned his birthday down to 30 April.

Working with the German physicist Wilhelm Weber in the 1930s, Gauss investigated magnetism and independently discovered **Kirchhoff**'s laws of electrical circuits. They also constructed the first electromechanical telegraph in 1833, creating a communications link between Göttingen's observatory and the town's physics institute.

Gauss also worked on a range of other physics topics including optics. In his nature, he was a perfectionist who was cautious about publishing his work; his personal diaries suggest he deserves credit for many mathematical discoveries that others made later, but published first. A standard unit for magnetic field strength was named the gauss in his honour.

ALEXANDER VON HUMBOLDT
(FRIEDRICH WILHELM HEINRICH ALEXANDER FREIHERR VON HUMBOLDT)
14 SEPTEMBER 1769 – 6 MAY 1859

German naturalist often considered a founder of modern geography

'I shall collect plants and fossils, and with the best of instruments make astronomic observations. Yet this is not the main purpose of my journey. I shall endeavour to find out how nature's forces act upon one another, and in what manner the geographic environment exerts its influence on animals and plants. In short, I must find out about the harmony in nature.'

From a letter written by Humboldt in 1799
(*The Life and Times of Alexander von Humboldt* by Helmut de Terra, 1955)

Born in Berlin, Humboldt studied in Frankfurt-an-der-Oder, Berlin and Göttingen, and at the Mining Academy in Freiburg. In 1799, he travelled to South America and spent five years studying various natural phenomena, including the links between volcanoes and fault lines, variations in the Earth's magnetic field near the equator, and the local flora and fauna.

In 1800, he mapped more than 2,700 km (1,700 miles) of the Orinoco River, describing animals such as the pink river dolphin. He returned via the US and during the expedition he collected around 60,000 plant specimens, many of which were unknown in Europe. Once back in Europe, Humboldt worked in Paris with **Gay-Lussac** on experiments to determine the composition of the air. He later served as a diplomat to the Prussian king Frederick William III.

In 1829, Humboldt was invited by Russian tsar Nicholas I to explore Siberia and Russian Asia. Late in life, he wrote up the knowledge he had gathered in *Kosmos*, a five-volume work that was the most accurate and comprehensive encyclopaedia of geology and geography of its time.

JOSEPH LOUIS GAY-LUSSAC
6 December 1778 – 9 May 1850

French chemist and physicist who formulated gas laws and explored new compounds

'If one were not animated with the desire to discover laws, they would escape the most enlightened attention'

Gay-Lussac quoted in *Gay-Lussac, Scientist and Bourgeois* by Maurice Crosland (1978)

Gay-Lussac was born at St Léonard, Haute-Vienne. From 1798, he studied at the École Polytechnique in Paris and from 1808 to 1832, he was professor of physics at the Sorbonne.

In 1802, he first published the 'law of volumes', which states that the volume of a gas increases linearly with its absolute temperature, providing the mass and pressure of the gas stays constant. However, Gay-Lussac credited unpublished work by Jacques Charles in the 1780s for the law, often known now as Charles's law. With **Humboldt**, Gay-Lussac showed in 1805 that one volume of oxygen combines with twice that volume of hydrogen to make water.

More generally, he showed that chemically combining volumes of gases always have simple ratios (such as one to one, or two to one), providing all volumes are measured at the same temperature and pressure. This gave support to **Dalton**'s atomic theory, although Dalton was reluctant to accept Gay-Lussac's experimental evidence. The issue would not be resolved until **Avogadro** developed a clearer picture of the nature of atoms and molecules.

Working with the chemist Louis Thenard, Gay-Lussac discovered and named the element boron, and he also recognized iodine as a new element. They created a wide range of new compounds, including dangerously reactive hydrogen fluoride. In experiments with potassium, Gay-Lussac suffered temporary blindness after triggering an explosion that destroyed his lab.

THOMAS YOUNG
13 June 1773 – 10 May 1829

English scientist who established the wave theory of light and helped decipher the Rosetta stone

'Scientific investigations are a sort of warfare, carried on in the closet or on the couch against all one's contemporaries and predecessors; I have often gained a signal victory when I have been half asleep, but more frequently found, on being thoroughly awake, that the enemy had still the advantage of me when I thought I had him fast in a corner – and all this, you see, keeps one alive'

Young quoted in *Thomas Young: Natural Philosopher* by Alexander Wood (1954)

Hailing from Milverton, Somerset, Young showed a remarkable talent for learning languages as a child, eventually learning more than a dozen. As a teenager he also mastered major scientific works including **Newton**'s *Principia*. In 1792, he began studying medicine at the University of Edinburgh and then at Göttingen, Germany, where he earned a medical degree in 1796. Having inherited a large sum of money from a relative, he set up his own medical practice in London, but he also continued studying at Cambridge University.

Young was professor of natural philosophy at the newly formed Royal Institution in London from 1801 to 1803. He later held various medical and scientific appointments, including serving as secretary to the Board of Longitude tasked with solving the problem of determining longitude at sea for navigation. His independent wealth also gave him ample opportunity to pursue his wide-ranging interests in experimental science.

Only in his early twenties, Young was elected a Fellow of the Royal Society for a 1794 paper on the human eye, which described how the ciliary muscle of the eye controls accommodation, the process of keeping an object in focus as its distance changes. A sufferer of astigmatism, he later showed that the condition stemmed from faulty curvature of the cornea.

But Young is best known for his work on the nature of light. At the time, this was a topic of hot debate. **Huygens** had introduced a wave theory of light, although Newton was convinced by the corpuscular theory, which held that light consists of a stream of discrete particles. Young suspected that the wave theory made more sense and he reasoned that if light does travel as waves, then these waves will demonstrate similar properties to ripples in water, which undeniably move as waves.

In a pool of water, waves that interact with each other interfere constructively and destructively, leaving distinctive interference patterns. In his famous double-slit experiment, Young showed that light did the same thing. Passing a narrow beam of light from a pinhole through two tiny slits, he found that he could create an interference pattern of bright and dark bands where the two beams had constructively and destructively interfered respectively.

Many scientists were very reluctant to accept Young's result, especially in England, given that it contradicted Newton's view. But it gained credibility when other scientists, including the French physicists Augustin-Jean Fresnel and Dominique-François Arago, repeated the experiment with the same results. In the 20th century, however, it became clear that both Young and Newton were correct. Light behaves as both waves and particles, depending on how you choose to observe it; this wave–particle duality is a central concept of quantum mechanics.

Young also related the colour of light to its wavelength, and estimated the approximate wavelengths of the seven colours that Newton had argued constituted a rainbow. He was first to suggest that the human eye has nerve fibres sensitive to three colours – red, green and blue. In 1817, he proposed that light waves are transverse (the undulations are perpendicular to the direction of travel), rather than longitudinal (with oscillations along the direction of travel). This explained how light can be polarized so that the waves tend to vibrate in the same plane.

In other work, Young experimented with liquids to study surface tension, and he related stress and strain in elastic media, showing that their stiffness can be characterized by a quantity now called Young's modulus, still an important concept in engineering. Late in life, Young branched out into Egyptology. He became the first person to successfully decipher a significant portion of hieroglyphics on the Rosetta Stone, discovered in the Nile Delta in 1799.

JEAN-BAPTISTE LAMARCK
(OFFICIALLY JEAN-BAPTISTE PIERRE ANTOINE DE MONET, CHEVALIER DE LA MARCK)
1 August 1744 – 18 December 1829

French naturalist and an early proponent of evolutionary theory

'The great age of the Earth will appear greater to man when he understands the origin of living organisms and the reasons for the gradual development and improvement of their organization. This antiquity will appear even greater when he realizes the length of time and the particular conditions which were necessary to bring all the living species into existence. This is particularly true since man is the latest result and present climax of this development, the ultimate limit of which, if it is ever reached, cannot be known.'

From Lamarck's *Hydrogéologie* (1802)

Born in Bazentin-le-Petit in northern France, Lamarck served in the army from 1761 to 1766 before studying medicine and botany in Paris. He published a long work on French flora in 1773 and in 1774 he became keeper of the royal garden in Paris. When it was transformed into the National Museum of Natural History in 1793, he was appointed professor of invertebrates.

Lamarck's work on fossils convinced him that new species develop from existing ones and this led him to promote an early theory of evolution. In his book *Philosophie Zoologique* (1809), he argued that species had to adapt to survive environmental change and that as species evolve, they gradually become more complex.

While this foreshadowed **Darwin**'s theory of evolution, Lamarck also promoted the erroneous notion that use or disuse of organs by individuals would make them develop further or shrink, and that these acquired features could be inherited. This idea was not seriously challenged in biology until around the 1880s.

JÖNS JACOB BERZELIUS
20 August 1779 – 7 August 1848

Swedish chemist considered one of the founders of modern chemistry

'When two oxidized bodies saturate one another, they always contain oxygen according to such proportions that the amount of oxygen in the body that goes to the positive pole in the circuit of the electrical pile is an integral multiple of the amount of oxygen in the other body, the one that tends toward the negative pole'

Berzelius describes evidence for the law of definite proportions
(*Enlightenment Science in the Romantic Era: The Chemistry of Berzelius and its Cultural Setting*, 1992)

Berzelius was born in Vaversunda and brought up by relatives after his parents died young. He studied medicine at the University of Uppsala from 1796 to 1801. After that he practised medicine in Stockholm, where a mining chemist Wilhelm Hisinger noticed his analytical talent and invited him to work in his laboratory. In 1807, he was appointed professor at the School of Surgery (later the Karolinska Institute) in Stockholm, but soon after that he abandoned medicine to focus on his chemistry research.

Around 1807, Berzelius began rigorous experiments to explain the chemical make-up of various compounds. He discovered so many examples of **Dalton**'s law of definite proportions – which states that chemical compounds are composed of atoms combined in fixed, whole number ratios – that chemists could no longer deny that Dalton was right. In 1828, Berzelius also compiled the first fairly accurate table of relative atomic weights for all elements known at the time.

Between 1803 and 1828, he discovered three new elements, cerium, selenium and thorium (although cerium had been independently isolated by Martin Heinrich Klaproth in Germany). He also introduced modern chemical formula notation, using symbols like C and O for carbon and oxygen, and he coined many modern chemical terms, including 'protein', 'catalysis' and 'isomer'.

JOHN DALTON
6 September 1766 – 27 July 1844

*English chemist and physicist who promoted
the atomic theory of matter*

'When we attempt to conceive the number of particles in
an atmosphere, it is somewhat like attempting to conceive
the number of stars in the Universe; we are confounded
with the thought. But if we limit the subject, by taking a
given volume of any gas, we seem persuaded that, let the
divisions be ever so minute, the number of particles must
be finite ... We might as well attempt to introduce a new
planet into the solar system, or to annihilate one already in
existence, as to create or destroy a particle of hydrogen. All
the changes we can produce consist in separating particles
that are in a state of cohesion or combination, and joining
those that were previously at a distance.'

From Dalton's *A New System of Chemical Philosophy* (1808)

**Dalton was born in Eaglesfield,
Cumberland** (now in Cumbria). The son
of a Quaker weaver, he became a school teacher
when he was only about 12 years old. In 1800, he
became secretary of the Manchester Literary and
Philosophical Society and he also privately taught
mathematics and chemistry. In 1817, he became
president of the Philosophical Society, an honorary
office that he held until his death.

He began his scientific research in his early twenties and his
interests were extremely broad, ranging from meteorology to natural
history. In meteorology, he linked the origin of trade winds to the Earth's
rotation and temperature variations (he was probably unaware that this
theory had already been proposed in 1735 by the English lawyer and
amateur meteorologist George Hadley). He also correctly identified that
the aurora borealis, or northern lights, must be linked to the Earth's
magnetic field.

Dalton was colour blind, and was the first person to investigate colour blindness in detail. 'Daltonism' became a common term for colour blindness. However, he incorrectly concluded that discolouration of liquid in the eyeball caused the disorder. But his most significant contribution was in chemistry. From his early studies on gases, he formulated the law of partial pressures (Dalton's law), which states that the total pressure of a mixture of gases that don't react with each other equals the sum of the pressures of the gases in the mixture. He also noted independently of **Gay-Lussac** and the French scientist Jacques Charles that a gas expands as its temperature rises.

Dalton devised a system of chemical symbols and proposed that a chemical combination of different elements occurs in simple numerical ratios by weight. He discovered butylene and correctly determined the composition of the solvent ether. But he is most famous for promoting the atomic theory – the idea that all elements are composed of tiny 'atoms' that are all alike and have the same atomic weight. He chose the term atoms because he recognized his theory mirrored that of **Democritus** more than 21 centuries earlier.

In Dalton's view, these particles are indestructible (which is true, disregarding nuclear reactions) and can only combine in whole-number ratios. Although that seems obvious today, in Dalton's time it was still a commonplace view that matter could be divided indefinitely. He also published a table of the relative atomic weights of the six elements hydrogen, oxygen, nitrogen, carbon, sulphur and phosphorus.

Dalton never married and had few friends, and as a lifelong Quaker, he lived a simple and unassuming life. But his scientific stature was so great that he was famous in his own lifetime. Some reports suggest that when he died in Manchester in 1844, around 40,000 people came to pay their respects.

HUMPHRY DAVY
17 December 1778 – 29 May 1829

British chemist and inventor who discovered new elements and compounds

'I began to respire 20 quarts of unmingled nitrous oxide. A thrilling, extending from the chest to the extremities, was almost immediately produced. I felt a sense of tangible extension high pleasurable in every limb; my visible impressions were dazzling and apparently magnified. I heard distinctly every sound in the room ... I lost all connection with external things ... As I recovered my former state of mind, I felt an inclination to communicate the discoveries I had made during the experiment. I endeavoured to recall the ideas – they were feeble and indistinct. One recollection of terms, however, presented itself, and with the most intense belief and prophetic manner, I exclaimed ... nothing exists but thoughts, the Universe is composed of impressions, ideas, pleasures and pains.'

Davy describes his experience of inhaling laughing gas
(quoted in *The Westminster Review*, 1824, vol 1)

Davy was born in Penzance, Cornwall. As a teenager he was apprenticed to a surgeon and apothecary in Penzance and he taught himself chemistry from textbooks, including **Lavoisier**'s major work. In 1798, he joined Bristol's Pneumatic Institution, which had been established to investigate possible medicinal powers of various gases. In 1801, he moved to the Royal Institution in London, where he became renowned for his entertaining lectures.

Working in Bristol, Davy prepared a wide range of gases and willingly inhaled them to test their effects. He inhaled pure hydrogen, almost suffocating in the process, as well as carbon dioxide. Most famously he prepared and noticed the inebriating effects of inhaling nitrous oxide and in 1800 published the results of his work in *Researches, Chemical and Philosophical; Chiefly Concerning Nitrous Oxide.* He even encouraged visitors to the institute to inhale it and coined the term 'laughing gas', but he didn't anticipate its medicinal use as an anaesthetic from the 1840s.

In other experiments, Davy came to the conclusion that heat was related to some kind of motion. In an 1806 lecture to the Royal Society, he predicted that electricity would be the key to splitting compounds into previously undiscovered elements. Many compounds, such as lime, were suspected to contain metallic elements that had never been isolated. Davy pioneered electrolysis using the voltaic pile (an early type of battery), and soon after that announced the isolation of the highly reactive alkali metals sodium and potassium. He discovered potassium in 1807 after electrolysing caustic potash (containing potassium, hydrogen and oxygen), and sodium by electrolysing molten sodium hydroxide.

He also isolated calcium by electrolysing a mixture of lime and mercury oxide. His other element discoveries included magnesium, boron and barium. Although **Scheele** was the first person to discover chlorine, Scheele thought it was a compound. Davy recognized correctly that it is an element and gave it its name, derived from the Greek word for green because of the colour of the gas.

Davy became president of the Royal Society in 1820, and for the last decade of his life, he spent most of his time travelling in Europe. In 1813, he visited Paris with his newly appointed assistant **Faraday** to collect a medal that Napoleon Bonaparte had awarded him for work on electrochemistry. There, he met **Gay-Lussac**, who asked him to identify a mysterious substance that another scientist had isolated. Davy identified it as a new element, now known as iodine. Later they travelled on to Italy, where Davy and Faraday proved that diamond is simply an isomer of carbon.

On returning to England, Davy invented the safety lamp for miners. Until then, frequent mining explosions had occurred due to miners using open flames that could ignite pockets of subterranean gas, such as methane. Davy came up with the idea of using iron gauze to shield the flame, so that methane burning inside the lamp would not spread outwards to create an uncontrolled explosion.

Although Davy's chemical research was forward thinking, he never accepted the atomic theory of his famous contemporary **Dalton**. Faraday went on to build on Davy's work and achieved even more fame in his own right. Davy was knighted in 1812. In 1819, he was awarded a baronetcy, at the time the highest honour a scientist had been awarded in Britain. In his will, he left money to finance an annual award for excellence in chemistry, which **Bunsen** and **Kirchhoff** received.

CHARLES BELL
12 November 1774 – 28 April 1842

Scottish surgeon and anatomist who pioneered neurophysiology research

'The cerebrum I consider as the grand organ by which the mind is united to the body. Into it all the nerves from the external organs of the senses enter; and from it all the nerves which are agents of the will pass out.'

From Bell's *Idea of a New Anatomy of the Brain* (1811)

Born in Edinburgh, Bell studied medicine at Edinburgh University, earning his degree in 1798. After that he became an assistant to his surgeon brother John Bell. He taught and researched anatomy at the city's Royal College of Surgeons and in 1904 moved to London, where he set up his own surgical practice. From 1812, he and his brother ran the Great Windmill Street School of Anatomy. He also served as a military surgeon before returning to Edinburgh University in 1836 as professor of surgery.

Bell is best known for his work on the nervous system. In *Idea of a New Anatomy of the Brain* (1811), he described nerves as bundles of fibres rather than single structures. He established that the sensory nerves could be traced from specific areas of the brain to different organs in the body. He later claimed that he was the first to recognize the difference between sensory nerves and motor nerves (in other words, sensory nerves transmit signals in one direction to the brain while motor nerves transmit signals from the brain to move muscles), although he didn't clearly document this. François Magendie, a French physiologist, is also credited with this discovery.

Bell also studied facial paralysis, linking the disease now known as Bell's palsy to damage to the facial nerve (cranial nerve VII, one of the 12 pairs of cranial nerves) that controls facial expression muscles. He was knighted in 1831.

HANS CHRISTIAN ØRSTED
(OR OERSTED)
14 August 1777 – 9 March 1851

Danish physicist and philosopher who discovered that electric currents create magnetic fields

'The agreement of this law with nature will be better seen by the repetition of experiments than by a long explanation'

Ørsted describes his discovery of magnetism arising from electric currents (from 'Experiments on the Effect of a Current of Electricity on the Magnetic Needle', 1820)

Ørsted was born in Rudkøbing, Langeland. He studied physics and philosophy at the University of Copenhagen, where he earned a doctorate in 1799 for a dissertation on the works of German philosopher Immanuel Kant. After travelling in Europe for three years on a scholarship, in 1806 he became professor at Copenhagen University, where he experimented in chemistry, electricity and acoustics. In 1829, he founded Denmark's College of Advanced Technology (now the Technical University of Denmark).

During a lecture in 1820, Ørsted noted that when an electric current from a battery was switched either on or off, the needle of a nearby compass was deflected. This confirmed that a changing electric current sets up a magnetic field. He went on to investigate the structure of the induced magnetic field and his work was influential in motivating **Ampère** to develop a mathematical theory of the magnetic forces between current-carrying conductors.

In chemistry, Ørsted succeeded in isolating aluminium for the first time in 1825. He also wrote poetry and works on philosophy. His brother was the politician Anders Sandøe Ørsted, who became prime minister of Denmark in 1853, and he was also a close friend of the author Hans Christian Andersen. The first Danish satellite, launched in 1999, was named after him.

CHARLES BABBAGE
26 December 1791 – 18 October 1871

English mathematician and inventor who developed the concept of a programmable computer

'The whole of the developments and operations of analysis are now capable of being executed by machinery ... As soon as an Analytical Engine exists, it will necessarily guide the future course of the science.'

From Babbage's autobiography *Passages from the Life of a Philosopher* (1864)

Born in London, Babbage developed an early interest in maths and studied at Cambridge University from 1810. With others, he founded the Analytic Society in 1815 to stimulate maths research and publicize developments from other European countries. In 1828, he was appointed Lucasian Professor of Mathematics at Cambridge, a post previously held by **Newton** and later by **Hawking**.

Babbage is best known for conceiving the idea of a mechanical computer. Although **Pascal** and **Leibniz** had designed primitive calculating machines, Babbage envisioned a more sophisticated one that could perform calculations using punched cards, store results (like the modern concept of computer memory) and perform further operations on them, then print out the results.

He began in 1822 by designing a 'difference engine', made to compute values of polynomial functions (functions such as $x^2 - 8x + 3$). Its design was unwieldy, composed of about 25,000 parts, and he never completed the machine, partly due to funding problems. He later designed an improved version. His 'analytical engine' design would have used punched cards and he was supported in this project by **Lovelace**, who created the world's first computer program for the machine.

Babbage's other inventions include an ophthalmoscope for eye examinations and a speedometer. With his wife Georgiana Whitmore, Babbage had eight children but only three survived into adulthood and his wife also died young; he suffered a mental breakdown following their deaths.

WILLIAM BUCKLAND
12 March 1784 – 14 August 1856

English geologist and palaeontologist who wrote the first detailed account of a fossil dinosaur

'Geology has shared the fate of other infant sciences, in being for a while considered hostile to revealed religion; so like them, when fully understood, it will be found a potent and consistent auxiliary to it, exalting our conviction of the power, and wisdom, and goodness of the Creator'

From Buckland's *Geology and Mineralogy considered with Reference to Natural Theology* (1836)

Buckland was born in Axminster, Devon, and enjoyed collecting fossils with his father as a child. He studied at the University of Oxford, earning a BA in 1805, and in 1808 he was ordained as a priest. At Oxford he attended lectures on mineralogy and chemistry, and in 1813 the university appointed him as a reader in mineralogy.

In 1824, Buckland published the first full account of a fossil dinosaur, which he named *Megalosaurus*, from the Stonesfield quarry in Oxfordshire. He estimated from the fragments that the animal was 12 m (40 ft) long. He also coined the term 'coprolites' for fossilized faeces, which he used to reconstruct ancient ecosystems.

Buckland recognized that fossil bones at Kirkdale Cave in Yorkshire were the remains of hyaenas and the animals they had eaten. Previously, he and others had suggested they were remains of tropical creatures swept up to the cave on the surging waters of the biblical flood. Though he was very keen to reconcile science with religion, Buckland gradually became persuaded by **Agassiz**'s glaciation theories and promoted them energetically in Britain.

He was also a renowned eccentric, who liked to parody the gait of extinct animals during his lectures. Famously, he also was obsessed with eating as many types of animal as possible, including bluebottles and moles.

ANDRÉ-MARIE AMPÈRE
20 January 1775 – 10 June 1836

French physicist and mathematician who created the science of electromagnetism

'For those of us who gather the fruits of their work without sharing their glory, I believe that above all we should make an effort to reduce to the smallest number possible the principles that should serve as a basis of all explications'

Ampère explains how scientists should try to discover the laws of nature (*André-Marie Ampère: Enlightenment and Electrodynamics* by James Hofmann, 2006)

Ampère was taught Latin by his father and developed an interest in maths and science, studying the works of **Bernoulli** and **Euler**. He gave private lessons in maths and chemistry from about 1796. From 1802 he held a series of professorships, becoming professor of mathematics at the École Polytechnique in Paris in 1809.

In September 1820, Ampère heard that **Ørsted** had shown that passing an electric current through a wire moved a magnetic needle nearby. Over the next few months, Ampère developed a mathematical theory describing this and showed that parallel wires are attracted to each other if electric currents passed through them in the same direction, while they repelled each other if the current directions were opposite. By 1825, he formulated Ampère's law, which states that the force between current-carrying wires is proportional to the product of the currents and inversely proportional to the square of the distance between them.

The standard unit of electric current, the ampere, is named after him. Biographers portray Ampère as prone to mood swings and emotionally scarred by the execution of his father, who was guillotined during the French Revolution, as well as the early death of his beloved first wife and a dysfunctional second marriage.

KARL ERNST VON BAER
29 February 1792 – 28 November 1876

German–Estonian naturalist who founded the science of embryology

'The general characters of the big group to which the embryo belongs appear in development earlier than the special characters ... The embryo of any given form, instead of passing through the state of other definite forms, on the contrary separates itself from them.'

From Baer's text *Über Entwickelungsgeschichte der Thiere*
(*History of the Evolution of Animals*, 1828)

Baer was born in Estonia to a German noble family. He studied medicine at the University of Dorpat (now Tartu, Estonia) then continued his studies at Berlin, Vienna and Würzburg. From 1817, he held professorships at Königsberg (now Kaliningrad in Russia), moving to the St Petersburg Academy of Sciences in 1834.

When Baer began his work on embryology, scientists often assumed that the follicles in the mammalian ovary were the eggs from which embryos develop. Baer showed that under the microscope, the true eggs in dogs are small yellow points inside the follicles and he concluded that all mammals, including people, develop from eggs. He also identified new structures within mammalian embryos, such as the notochord and neural folds that go on to form the backbone and central nervous system.

Baer recognized that in embryo development, the general features of a large group of animals appear earlier than specialized ones, a rule now known as Baer's first law. This made preformation – the idea that animals develop from fully formed but miniature versions of themselves – completely untenable. He also dispelled any notion that features of different animals develop in a single embryo through stages; an embryonic human arm or leg doesn't form by passing through a wing or a hoof stage, for instance. Instead, general limb buds form in animal embryos before diverging into distinct limbs.

JUSTUS VON LIEBIG
12 May 1803 – 18 April 1873

German chemist who promoted chemistry education and identified chemical isomers

'Chemistry, as an independent science, offers one of the most powerful means towards the attainment of a higher mental cultivation; that the study of chemistry is profitable, not only inasmuch as it promotes the material interests of mankind, but also because it furnishes us with insight into those wonders of creation which immediately surround us, and with which our existence, life, and development, are most closely connected'

From Liebig's *Familiar letters on chemistry in its relations to physiology, dietetics, agriculture, commerce and political economy* (1851)

Liebig was born in Darmstadt. He studied at the University of Bonn from 1820 and at the University of Erlangen, where he earned a doctorate. In 1822, he moved to Paris to work in **Gay-Lussac**'s laboratory, and in 1824 he was appointed on **Humboldt**'s recommendation as professor at the University of Giessen, where he worked for nearly 30 years. In 1852, he moved to Munich where he remained for the rest of his life.

At Giessen, Liebig set up a thriving teaching laboratory for organic chemistry. In early work there, he showed that the fulminate ion (CNO^-) has the same chemical formula as cyanate (OCN^-), introducing the concept of 'isomerism' – that two chemical compounds can contain the same set of atoms yet have very different properties because the atoms are arranged differently. He also introduced an efficient technique for analysing organic compounds by burning them, oxidizing them to carbon dioxide and water, and weighing the products.

Liebig was also influential in changing agricultural practice, carrying out early experiments with synthetic fertilizers that eventually led to their widespread use. He also established that carbohydrates and fats in our diets serve as the body's fuel, rather than carbon and hydrogen, as **Lavoisier** had suggested.

AMEDEO AVOGADRO
(LORENZO ROMANO AMEDEO CARLO AVOGADRO)
9 AUGUST 1776 – 9 JULY 1856

Italian scientist best known for his work on molecular theory

'To Avogadro ... the molecules and atoms considered in this great
theory were real objects: they were thought of the same way as one
thinks of tables and chairs'

John Bradley describes Avogadro's views in a report in the
British Journal for the Philosophy of Science (1955, vol 6)

Avogadro studied ecclesiastical law but also showed an
interest in maths and natural philosophy, which he studied from around
1800. In 1809, he was appointed professor of natural philosophy at the
college of Vercelli. In 1820, he became professor of mathematical physics
at the University of Turin.

After studies of electricity, Avogadro focused on gases. He knew that
Gay-Lussac had shown that all gases expand to the same extent as the
temperature rises, so he reasoned that this must be because they contain
the same number of particles per unit volume (at one given temperature).
He realized that the particles could be molecules, a term that he coined, as
well as atoms, which he called 'elementary molecules'.

He went on to reason that when water is electrolyzed, liberating
hydrogen and oxygen gas at separate electrodes, the reason that there's
twice as much hydrogen as oxygen by volume is that water molecules
contain two hydrogen atoms for every oxygen atom. The number
of molecules in a mole (one gram molecular weight) is today called
Avogadro's number, which is equal to 6.022×10^{23} molecules per gram
mole. For instance, oxygen has a molecular mass of 32, so 32 grams of
oxygen contains Avogadro's number of molecules. Confusion between
atoms and molecules persisted, however, and Avogadro's theory wasn't
generally accepted until several years after his death.

ROBERT BROWN
21 December 1773 – 10 June 1858

Scottish botanist and pioneer of microscopy who first observed Brownian motion

'These motions were such as to satisfy me, after frequently repeated observation, that they arose neither from currents in the fluid, nor from its gradual evaporation, but belonged to the particle itself'

From a report by Brown in *London and Edinburgh Philosophical Magazine and Journal of Science* (1828)

Born in Montrose, Brown studied medicine at the University of Edinburgh and became an army medical officer stationed in Ireland for five years. But his chief passion was botany. He met **Banks** during a trip to London in 1798, and Banks recommended him as ship's naturalist for Matthew Flinders' expedition to the Cape of Good Hope and Australia, which set sail in 1801. He also worked as a librarian for Banks and later was keeper of the botanical department at the British Museum.

During the voyage, Brown collected nearly 4,000 plant specimens and he published a major systematic account of the Australian flora in 1810. He observed the cell nucleus, which had been observed by earlier microscopists, but Brown gave it the name nucleus for the first time. And in 1827, he made the first detailed studies of what is now called Brownian motion. Brownian motion describes the jittery, random motion of relatively large particles suspended in a fluid or gas, such as smoke particles in air.

Brown noticed that small particles ejected by pollen grains in water jiggled about, following a zigzag path. He confirmed that the effect also occurred for inorganic matter particles, ruling out the idea that it was life-related. In 1905, **Einstein** showed that Brownian motion can be predicted mathematically by assuming the suspended particles are constantly bumped by fluid molecules moving due to their own thermal energy, indirectly proving that molecules exist.

FRIEDRICH WÖHLER
31 July 1800 – 23 September 1882

German chemist who isolated chemical elements and radically changed attitudes about the nature of organic compounds

'This investigation has yielded an unanticipated result that reaction of cyanic acid with ammonia gives urea, a noteworthy result in as much as it provides an example of the artificial production of an organic, indeed a so-called animal, substance from inorganic substances'

Wöhler quoted in *Annalen der Physik und Chemie* (1828, vol 88)

Wöhler studied medicine in Heidelberg and taught chemistry from 1825 at the polytechnic schools of Berlin and Kassel. In 1836, he became professor of chemistry at the University of Göttingen, where he remained until his death.

In 1828, Wöhler accidentally synthesized urea, which has the chemical formula $CO(NH_2)_2$. Previously, **Berzelius** and others had promoted the idea that 'organic' compounds formed only from carbon, oxygen, hydrogen and nitrogen were always products of living organisms (urea is a major component of urine). It was considered impossible to create these compounds from inorganic materials in the lab, so many scientists believed some mysterious 'vital force' was necessary to explain their creation. Wöhler's synthesis of urea and later synthesis of other natural products eventually laid the vitalism theory to rest.

Wöhler also discovered or co-discovered and isolated several elements and compounds including yttrium, beryllium, titanium, silicon, aluminium and silicon nitride. A keen collector of meteorites, he showed that some of the stony ones contain organic matter. Much of his work was done in collaboration with **Liebig**, and they are often considered to have laid much of the foundations of modern organic chemistry.

JOSEPH HENRY
17 December 1797 – 13 May 1878

American scientist who discovered the electromagnetic phenomenon of self-inductance

'The seeds of great discoveries are constantly floating around us, but they only take root in the minds well prepared to receive them'

Quote commonly attributed to Joseph Henry

Henry was born in Albany, New York, to poor Scottish immigrants. At the age of 13 he became an apprentice to a watchmaker but soon after he stumbled on a book of scientific lectures and became hooked. From 1819 he was given free tuition at the Albany Academy, and although he intended to become a medical doctor, he changed his plans when offered a job as an assistant engineer for road building in 1824. In 1826, he was appointed professor at the Albany Academy and two decades later he was elected first secretary of the Smithsonian Institution of Washington.

Learning of earlier work on electromagnets, Henry believed he could improve on them by properly insulating their many coils to prevent short circuits. By 1831, he had developed an electromagnet so powerful it could lift more than a tonne of iron off the ground using an ordinary battery. He also effectively invented the telegraph, by sending a current over a mile of wire to activate an electromagnet that made a bell strike, although it wasn't a practical technique until improved on and commercialized by American artist-turned-inventor Samuel Morse.

In 1830, Henry discovered electrical inductance, in which an electric current flowing in a metal coil sets up a magnetic field which in turn sets up a current in a nearby coil (although **Faraday** published this discovery first). Henry also discovered self-inductance, when a current flowing in a coil induces a current in the same coil. The metric unit of inductance, the henry, is named after him.

CHARLES LYELL
14 November 1797 – 22 February 1875

Scottish geologist who popularized the notion of gradual change shaping the Earth's surface

'We find in certain localities subterranean deposits of coal, consisting of vegetable matter, formerly drifted into seas and lakes. These seas and lakes have since been filled up, the lands whereon the forests grew have disappeared or changed their form, the rivers and currents which floated the vegetable masses can no longer be traced, and the plants belonged to species which for ages have passed away from the surface of our planet.'

From Lyell's *Principles of Geology* (1830 – 1833)

Lyell was born in Kinnordy, north of Dundee. From 1816 he studied classics and law at Oxford University, where he attended geology lectures by **Buckland**. After graduation he became a lawyer but continued his geological interests as Secretary to the Geological Society of London, and he gave up law in 1828. After travelling in Europe for geological research, he published the first volume of his famous work *Principles of Geology* in 1830, and he was appointed professor of geology at King's College, London, in 1831.

At that time, it was still common for geologists to suggest that one or more worldwide 'catastrophes' had shaped the Earth's surface. Lyell's work led to general acceptance of **Hutton's** concept of uniformitarianism, the notion that geological formations are the result of slow and gradual processes that are still going on today, and are therefore observable. Lyell argued that these processes changed the Earth over millions of years, as opposed to thousands of years implied by Biblical interpretations.

His three-volume *Principles of Geology* persuasively argued his case and proved a stunning success, being printed in 12 editions by 1875. He was knighted in 1848.

22 September 1791 – 25 August 1867

English natural philosopher who established the concept of the electromagnetic field

'Electricity is often called wonderful, beautiful; but it is so only in common with the other forces of nature. The beauty of electricity or of any other force is not that the power is mysterious, and unexpected, touching every sense at unawares in turn, but that it is under law, and that the taught intellect can even now govern it largely. The human mind is placed above, and not beneath it, and it is in such a point of view that the mental education afforded by science is rendered super-eminent in dignity, in practical application and utility; for by enabling the mind to apply the natural power through law, it conveys the gifts of God to man.'

From Faraday's lecture notes
(quoted in *The Life and Letters of Faraday* by Bence Jones, 1870)

Faraday was born in Newington, Surrey (now part of the London borough of Southwark). He was apprenticed to a London bookbinder in 1805 and there he taught himself chemistry and physics by reading textbooks, including work by **Lavoisier**. In 1812, a business customer gave Faraday tickets for **Davy**'s lectures at the Royal Institution and this was to change the course of his life. Faraday wrote extensive notes on the lectures, which he sent to **Banks**, who was then president of the Royal Society, and also to Davy himself. In 1813, Davy appointed him as his assistant at the Royal Institution in London and in 1833, Faraday became Fullerian Professor of Chemistry there.

In 1820, **Ørsted** had discovered that a changing electric current produces a magnetic field. Faraday was among many scientists who were intrigued to investigate this, and in doing so, he invented the electric motor. Using a wire, magnet and chemical battery, he showed that current flow in the wire would make it spin around the magnet. Faraday's intuition was that just as Ørsted had created a magnetic field from an electric current, he could create an electric current from a magnetic field. Sure enough, his experiments showed that a magnetic field, itself generated by a changing electric current in a coil, induced a current in a second coil – an effect now known as induction. Induction was independently discovered by **Henry** in the US.

Faraday recognized that it might be possible to generate continuous electric current by adapting an experiment by the French scientist, Dominique Arago. In 1824, Arago had discovered that a rotating copper disc deflected a magnetic needle. Faraday realized this was due to induction and by reversing the experiment – spinning a copper disc between the poles of a permanent magnet – he set up an electric current in the disk that could be passed through a wire. Effectively, he had invented the electric generator, although it was not until many decades later that electric generators based on this principle became practical. Faraday also clearly established that there was a deep underlying relationship between magnetism and light.

Faraday is credited with introducing the concepts of magnetic lines of force and of force 'fields' that extend through space and can be produced by magnets, electric charge or mass, in the case of gravity. Lacking advanced maths skills, he did not develop these concepts in any rigorous theoretical way. That was left to **Maxwell**, who later developed his four famous equations describing electromagnetism, one of them now expressing what's known as Faraday's law of induction.

In work on static electricity, Faraday showed that the electric field inside a closed conductor is zero, because the charges redistribute so that the interior fields cancel. A 'Faraday cage', typically a closed mesh of metal, is often used to protect equipment from electrical damage. His other discoveries include the aromatic organic compound benzene, which he called 'bicarburet of hydrogen'. He also experimented with liquefying gases such as chlorine.

Davy became bitter about his assistant's meteoric rise and when Faraday was nominated to become a member of the Royal Society in 1824, Davy cast the only negative vote. Although Faraday valued membership of the society, he turned down an invitation to be its president and, with a dislike of vanity, he also turned down a knighthood. He was asked to study the feasibility of using poison gas on the battlefield during the Crimean War in the 1850s, and responded that while it was feasible, he personally wanted nothing to do with it. The standard international unit of capacitance (the farad) is named in his honour.

(JEAN) LOUIS (RODOLPHE) AGASSIZ
28 May 1807 – 14 December 1873

Swiss–American palaeontologist, glaciologist and biologist who first suggested Earth had suffered ice ages

'The glacier was God's great plough'

From Agassiz's *Geological Sketches* (1866)

Agassiz was born in Fribourg, Switzerland, and studied medicine at universities in Germany, where he also developed his interest in natural history. Afterwards he worked with **Humboldt** in Paris and in 1832 he became professor of natural history back in Switzerland at the University of Neuchâtel. He moved in 1846 to the US, where he was appointed professor of zoology and geology at Harvard University in 1847.

Between 1833 to 1843, Agassiz published five volumes (*Recherches sur les Poissons Fossiles*, or *Research on Fossil Fish*) in which he described and classified more than 1,700 fossil fish. In 1837, he also became the first person to propose on a scientific footing that the Earth was subjected to ice ages in the past. Some scientists had speculated that some haphazard collections of rocks in the Alps had been deposited there by moving glaciers. To investigate this, Agassiz arranged for a hut to be built on a glacier in the Bernese Alps and lived there to monitor its movement.

Sure enough, he found that the glacier moved. In 1840, he published *Etudes sur les Glaciers* (*Study on Glaciers*), in which he described how glaciers erode rocks as they move and their influence on Alpine landscapes. His dramatic conclusion, following further observations in Britain, was that ice sheets had once extended down over much of northern Europe during bitterly cold ice ages.

Throughout his working life, Agassiz opposed **Darwin**'s theories on evolution. He considered different races of people to be different species with separate origins and believed that the Biblical Book of Genesis described only the origins of white people, although he did reject racism.

FRIEDRICH WILHELM BESSEL
22 July 1784 – 17 March 1846

German mathematician and astronomer who made the first accurate measurement of the distance to a star beyond the Sun

'Every astronomical instrument is made twice. First in the shop of the artisan who makes it as expertly as he can. Second in the astronomical observatory where the astronomer carefully works out its errors and, by meticulously allowing for them, makes the instrument into a more nearly perfect device than it had been when it left the artisan's hands.'

Bessel quoted in *Eyes on the Universe* by Isaac Asimov (1975)

Bessel left school aged 14 and worked for an import–export business in Bremen, becoming the company's accountant. The business's shipping concerns led him to study maths and astronomy as a route to solving navigation problems, and he was later appointed as an assistant at Lilienthal Observatory near Bremen, where he catalogued the positions of thousands of stars. In 1810, King Frederick William III of Prussia appointed him as director of the Königsberg Observatory.

Bessel introduced a powerful new method for predicting solar eclipses and he also generalized what are now known as the Bessel functions, solutions of a differential equation that have important applications in physics (they were first discovered by **Bernoulli**). However, Bessel is most famous for being first to measure the distance to a star using the technique of 'parallax'.

Over six months, the Earth moves half way round its orbit of the Sun, so the angular position of a star viewed from the two vantage points is slightly different. In 1838, Bessel showed that the angular position of the star 61 Cygni changed by 0.628 arcseconds (where an arcsecond is 1/360th of a degree) over the course of six months. Knowing the diameter of the Earth's orbit, he could then calculate that the star lies about 10.4 light years away.

CHRISTIAN JOHANN DOPPLER
29 NOVEMBER 1803 – 17 MARCH 1853

Austrian mathematician and physicist who discovered the 'Doppler effect', which changes the pitch of sound from moving sources or changes their colour

'We know from general experience that a ship of moderately deep draught which is steering toward the oncoming waves has to receive, in the same period of time, more waves and with a greater impact than one which is not moving or is even moving along in the direction. If this is valid for the waves of water, then why should it not also be applied with necessary modification to air and ether waves?'

Quoted in 'Christian Johann Doppler: the man behind the effect' by A. Roguin (2002, vol 75)

The son of a stonemason, Doppler was born and raised in Salzburg. He studied maths at the Vienna Polytechnic Institute (now Vienna University of Technology) then studied maths, mechanics and astronomy at the University of Vienna, where he later stayed on as a research assistant. After various teaching posts he was appointed professor at the Technical Institute of Prague in 1841.

The following year, Doppler announced the idea that would make him famous. He predicted that the frequency of a star's light – which also determines its colour – should be dependent on the star's motion relative to observers on Earth. When the star moves towards Earth, each successive light wave emitted comes from a position closer to the observer, giving it a shorter travel time. Effectively, the waves bunch together creating an increase in frequency. Conversely, when the wave source moves away, successive waves are emitted from greater distance. The waves stretch out, making their frequency drop.

This means green light would appear more blue when approaching Earth at very high speed and more red when receding. He had no means of testing this, but he also recognized it should happen for sound waves too. A sound source moving towards an observer should be higher in pitch as waves bunch together, while the same sound source would be lower in pitch when receding. (This explains why the siren of a fire engine sounds high-pitched as it comes towards us and lower-pitched after the vehicle has driven past.)

To test his theory, Doppler arranged for his colleague, the Dutch scientist Christophorus Buys Ballot, to perform one of the quirkiest experiments on record. In 1845, Buys Ballot arranged for a train to pull an open car full of trumpeters back and forth along the train line between Utrecht and Amsterdam while stationary trumpeters played the same note on the track side. The train locomotive was capable of reaching about 64 km/h (40 mph), considered very fast at the time.

Doppler had predicted that pitch changes due to the train approaching or receding would make the sounds from the two groups of musicians dissonant, and sure enough, it was. Later, the French physicist Armand Hippolyte Louis Fizeau generalized Doppler's work for light and showed in 1848 that the Doppler effect is indeed responsible for changes in starlight (shifts in the characteristic spectral patterns of starlight had by now been measured).

Doppler later held teaching posts in Vienna and died aged 50 of tuberculosis in Venice, where he had hoped to convalesce. The effect he described today underpins an amazingly diverse range of techniques, including methods of weather forecasting and navigation, detecting alien planets circling stars beyond the Sun, and diagnosing cardio-vascular disease.

MATTHEW FONTAINE MAURY
14 January 1806 – 1 February 1873

American oceanographer who improved marine navigation

'The immediate result of Mr Maury's labour is, that ocean voyages
under sail are shortened from ten to twenty per cent; and if this result
be followed out to its legitimate consequences, who can undertake to
prescribe a limit to the benefits they confer? Who will undertake to
estimate the mere pecuniary saving, to the navigating interests, in the
decreased expenditure for outfits, provisioning and manning ships,
the decrease in ocean risks, not only to ships and cargoes, but to lives
of seamen and passengers?'

From a report on Lieutenant Matthew Fontaine Maury's services (1855)

Maury was born in Fredericksburg, Virginia, and joined
the US Navy as a midshipman in 1825 at the age of 19. When a leg injury
put an end to his sea duty, he became superintendent of the Depot of
Charts and Instruments, and director of the US Naval Observatory in
Washington DC. With the outbreak of the Civil War, he resigned to join
the Confederacy, for which he acquired ships. He later held a teaching post
at the Virginia Military Institute in Lexington.

On joining the navy, Maury enthusiastically began scientific studies of
the oceans and methods of navigation. His *Physical Geography of the Sea*
(1855) was the first extensive and comprehensive book on oceanography.
At the Naval Observatory, he studied thousands of ships' logs to synthesize
information on ocean currents and the effects of adverse winds on a ship's
course, and this allowed him to publish charts that captains could use to
exploit ocean currents and winds to their advantage, drastically cutting
sailing times.

After analysing migration patterns of whales, Maury also predicted the
existence of the Northwest Passage, a sea route through the Arctic Ocean
connecting the Atlantic and Pacific oceans. He was a founding member of
the American Association for the Advancement of Science in 1848. From
1849, he used research vessels to map ocean temperatures and collect
samples from the seafloor.

JAMES PRESCOTT JOULE
24 December 1818 – 11 October 1889

English physicist who studied the nature of heat and mechanical work

'The Earth in its rapid motion round the Sun possesses
a degree of living force so vast that, if turned into the
equivalent of heat, its temperature would be rendered at
least one thousand times greater than that of red-hot iron,
and the globe on which we tread would in all probability
be rendered equal in brightness to the Sun itself'

From Joule's *On Matter, Living Force, and Heat* (1847)

Joule was home schooled in Salford until
1834 when he and his brother began studies with
Dalton. From an early age, he indulged in eccentric
experiments, flying kites in thunderstorms and
giving a family servant increasingly strong electric
shocks to find out her sensations. He later became a
manager of the family brewing business until it was
sold in 1854.

In 1840, Joule showed that an electrical current flowing through a wire
makes it hot, and that the heat is proportional to the product of the wire's
resistance, the square of the current and its duration (Joule's law). By 1843,
he had proven in experiments that the heat is actually generated inside
the wire rather than being transferred from equipment connected to it,
breaking from the traditional 'caloric theory' that heat couldn't be created
nor destroyed.

Joule went on to estimate the mechanical equivalent of heat and in
work with **Kelvin**, he showed that when a gas expands without doing
external work, the temperature of the gas falls. A standard unit of energy,
the joule, is named after him.

ADA LOVELACE
(AUGUSTA ADA KING, NÉE BYRON,
COUNTESS OF LOVELACE)
10 December 1815 – 27 November 1852

*English mathematician often considered to
be the first computer programmer*

'The Analytical Engine weaves algebraic patterns just as
the Jacquard loom weaves flowers and leaves'

From Lovelace's notes in *Sketch of the Analytical Engine* by Luigi Menabrea

Ada Augusta Byron, whose father was the poet Lord Byron, was born in London. She was privately educated in maths and science and in 1833 was introduced to **Babbage**. In 1835, she married William King, who later became the first Earl of Lovelace.

Lovelace translated a description of Babbage's calculating machine, the punched-card-input 'Analytical Engine', by the Italian mathematician Luigi Menabrea. She added extensive notes of her own, including a method for calculating a sequence of Bernoulli numbers, a formula-generated sequence of numbers that alternate between positive and negative. The Analytical Engine would have correctly generated this number sequence had it been built. For this work, Lovelace is often considered to have created the world's first computer program. She also predicted that the Analytical Engine might be used to compose music and generate graphics.

Impressed with her work, Babbage described Lovelace as 'the Enchantress of Numbers'. She died young aged 36, after her physicians prescribed bloodletting when she developed cancer. The computer language Ada, designed between 1977 and 1983 for the US Department of Defense, was named in her honour.

GUSTAV ROBERT KIRCHHOFF
12 March 1824 – 17 October 1887

German physicist famous for work on electrical circuits and radiation

'Look here, I have succeeded at last in fetching some gold from the Sun'

Kirchhoff's remark to his banker after winning gold for researching elements in the Sun; his banker had questioned the value of his work (*A memoir of Gustav Robert Kirchhoff*, by Robert Von Helmholtz, 1890)

Kirchhoff was born in Königsberg, East Prussia (now Kaliningrad, Russia). He graduated from the Albertus University of Königsberg in 1847. From 1854, he was professor at the University of Heidelberg and in 1875 he accepted the first professorship specifically dedicated to theoretical physics at Berlin University.

In 1845, while still a student, Kirchhoff formulated his famous electrical circuit laws. He showed that current is conserved, so that the total amount of electrical current flowing into a junction is the same as the total coming out. His voltage law says that the algebraic sum of the voltage (potential) differences in any loop must add to zero.

In Heidelberg, Kirchhoff worked on spectroscopy with **Bunsen** and they jointly discovered the elements caesium and rubidium in 1861. They made the key link that the spectral lines emitted by a hot gas occurred at the same wavelength, in modern parlance, as the dark absorption lines observed when incandescent light shines through the same gas heated to the same temperature. Kirchhoff recognized that the dark spectral lines in the solar spectrum could therefore reveal its composition, and he demonstrated the existence in the Sun of many elements already isolated on Earth.

HERMANN LUDWIG FERDINAND VON HELMHOLTZ
31 August 1821 – 8 September 1894

German physicist and physician who did pioneering studies in physics and physiology

'I have been able to solve a few problems of mathematical physics on which the greatest mathematicians since Euler have struggled in vain ... But the pride I might have held in my conclusions was perceptibly lessened by the fact that I knew that the solution of these problems had almost always come to me as the gradual generalization of favourable examples, by a series of fortunate conjectures, after many errors. I am fain to compare myself with a wanderer on the mountains who, not knowing the path, climbs slowly and painfully upwards and often has to retrace his steps because he can go no further ... when he reaches the summit he finds to his shame that there is a royal road by which he might have ascended, had he only the wits to find the right approach to it.'

Helmholtz quoted in *Hermann von Helmholtz* by Leo Koenigsberger (1906)

Born in Potsdam, Helmholtz developed an early interest in physics but his family couldn't fund his university education so he instead studied medicine in Berlin with state funding on condition that he served as a doctor in the Prussian army afterwards. During his service as an army surgeon from 1843 to 1848, he conducted scientific research in his spare time.

In 1849, he was appointed as associate professor of physiology at the University of Königsberg (now Kaliningrad in Russia) and in 1855 he became professor at the University of Bonn. Later he accepted professorships in physiology and then physics at the universities of Heidelberg and Berlin.

In early work on the physiology of the eye, Helmholtz invented the ophthalmoscope for examining the eye's interior independently of **Babbage**. He also revived and developed **Young**'s three-colour theory of vision. In studies of the ear, he explained pitch detection in the inner ear by the cochlea and showed how the quality of a tone depends on the number of overtones (harmonics) and their relative intensities.

Helmholtz also made important contributions to neurophysiology. At the time, it was a common perception that nerve impulses moved so quickly that they would be impossible to measure. But following experiments on frog muscles, Helmholtz reported transmission speeds of roughly 30 m/s (70 mph).

However, he is best known for his work in physics, in which he developed the concept of conservation of energy with the most detailed argument to date. **Liebig** had already focused attention on the question of whether an animal's metabolism produces all its heat and mechanical energy. If so, then the theory of vitalism – that the functions of a living organism are driven by some kind of 'vital principle' beyond biochemical reactions – would be superfluous. Through experiments on frog muscles using electrical currents, Helmholtz showed that the heat generated by muscles is easily accounted for by metabolism and muscular action.

Some proponents of vitalism had argued that the vital force, whether mechanical or not, might be an inexhaustible force that could power a perpetual motion machine. Helmholtz stressed that living organisms and machinery simply consume the power of natural forces and reasoned that energy must be conserved, and he argued that heat and work can be substituted for each other quantitatively.

He also investigated the possible energy source of the Sun. At that time, the only reasonable source of energy seemed to be gravitation. As **Laplace** had argued in his nebular hypothesis, the Sun started out as a vast gas cloud that gradually contracted. Helmholtz reasoned that the Sun's energy could come from the kinetic energy the particles attained falling towards the centre of the cloud. From the amount of radiation emitted by the Sun, he calculated the necessary contraction rate and working backwards, he showed that the Sun would have been big enough to engulf the Earth 25 million years ago – which he argued must have been the maximum age of the Earth. Today, it is clear that the Sun is in fact powered by nuclear reactions in its core and that the Earth is roughly 4.5 billion years old.

Helmholtz also defined the principles of fluid flow and developed important principles that underpin thermodynamics. In Heidelberg, he worked with **Kirchhoff** and **Bunsen**, and during his career he mentored several notable students, including **Wundt** and Heinrich Hertz, who clarified and expanded the electromagnetic theory of light developed by **Maxwell**. He was world-renowned in his lifetime and is remembered as one of the most versatile scientists of the 19th century.

JOHANN GOTTFRIED GALLE
9 June 1812 – 10 July 1910

German astronomer who discovered Neptune

'The planet whose position you have pointed out actually
exists. The same day that I received your letter, I found
a star of the eighth magnitude which was not shown on
the excellent chart ... published by the Royal Academy
of Berlin. The observations made the following day
determined that this was the sought-for planet.'

Letter from Galle to French mathematician Urbain Le Verrier, 1846
(*From Eudoxus to Einstein* by Christopher Linton, 2004)

Galle was born in Pabsthaus, Saxony. He studied at the
University of Berlin from 1830 and worked as an assistant at the Berlin
Observatory, where he eventually became director. In 1851, he was
appointed professor at the University of Breslau (now Wrocław in Poland).

Galle correctly proposed that astronomers could measure the distance
to the Sun more accurately than before by measuring the 'parallax' of
asteroids to see how their position changed over the course of six months,
or half an Earth orbit. That allowed the distance to the asteroids, and the
Sun, to be estimated. But he is best known for discovering the planet
Neptune. Observations of Uranus had shown inconsistencies in its
motion that suggested the gravity of another large planet was disturbing
Uranus's orbit. The French mathematician Urbain Le Verrier calculated
the likely position of this planet, and using this prediction, Galle observed
the expected planet on 23 September 1846.

The following night, he found that it had moved against the
background of the fixed stars confirming it must be a planet in the solar
system. Although Neptune had previously been observed by **Galileo** and
probably many others since the invention of the telescope, Galle was the
first to recognize what it actually was. A distinctive 'smiley face' crater on
Mars is named Galle in his honour.

(JEAN BERNARD) LÉON FOUCAULT
18 September 1819 – 11 February 1868

French physicist who invented the Foucault pendulum that demonstrated the Earth's rotation

'One feels, one sees it born and grow steadily; and it is not in one's power to either hasten it or slow it down. Any person, brought into the presence of this fact, stops for a few moments and remains pensive and silent; and then generally leaves, carrying with him forever a sharper, keener sense of our incessant motion through space.'

Foucault describes the power of his pendulum demonstration (quoted in *Pendulum: Léon Foucault and the Triumph of Science* by Amir Aczel, 2003)

Foucault had little formal education but was hired as an experimental assistant by a wealthy French physicist, Hippolyte Fizeau. With Fizeau, Foucault made early measurements of the speed of light using a system of mirrors. By 1862, he had improved Fizeau's apparatus to make the most accurate light speed measurement on record, within about 0.6 per cent of the modern value.

But Foucault is best known for his pendulum experiments. In 1851, he suspended a heavy iron ball from inside the dome of the Paris Panthéon on a wire about 67 m (222 ft) long. A pointer on the base of the ball was in contact with sand underneath, and when the pendulum was made to swing in wide sweeps, the sand recorded the way the plane of the pendulum's swing drifted round due to the movement of the Panthéon's floor.

This was the first experimental demonstration showing that the Earth rotates. Foucault also demonstrated this using the gyroscope, a device that he named but did not invent. He died of some kind of progressive paralysis disease at the young age of 48.

WILLIAM THOMSON, LORD KELVIN
26 June 1824 – 17 December 1907

British physicist and engineer regarded as a founder of modern physics

'I often say that when you can measure what you are speaking about and express it in numbers you know something about it; but when you cannot measure it, when you cannot express it in numbers, your knowledge is of a meagre and unsatisfactory kind: it may be the beginning of knowledge, but you have scarcely, in your thoughts, advanced to the stage of science'

From Kelvin's lecture delivered at the Institution of Civil Engineers (1883)

 **Thomson's father was a maths** and engineering teacher in Belfast, and Thomson was home schooled until his family moved to Glasgow in 1833. From 1834 he studied at the University of Glasgow (it was normal at the time for the university to teach children as young as 10 years old) and he excelled in both classics and sciences.

From 1841 he studied at Cambridge University, graduating in 1845, and the following year he became professor of natural philosophy at the University of Glasgow, aged only 22. He remained there for more than half a century. In 1892, he was made a peer and adopted the title Baron Kelvin of Largs (he's most commonly referred to now as Lord Kelvin).

With **Faraday** and **Maxwell**, Kelvin built the foundations for the science of electromagnetism. He mathematically developed Faraday's theory of induction, in which a changing current flowing through a coil induces a second electric current in a nearby coil. Kelvin's work demonstrated that the effect is transmitted via the intervening medium, rather than being some mysterious 'action at a distance'. He also introduced vectors to represent quantities such as magnetic induction and magnetic force.

Working with **Joule** on thermodynamics, Kelvin firmly established the notion that energy is conserved. He realized that as a body continually cools, it must reach a point at which no further heat energy can be removed, the point of absolute zero. Based on that he proposed the absolute temperature scale (now the kelvin scale) which uses the degree Celsius as its unit increment; 0 K corresponds to −273.16°C.

Kelvin also formulated a version of the second law of thermodynamics, which is today normally framed in terms of 'entropy'. Because energy varies in its quality or ability to do useful work, the entropy of an isolated system – a measure of the energy input that doesn't do mechanical work – always increases. Influenced by **Davy** and Joule, he backed the theory that heat is a form of motion.

But it was his work on telegraphy that finally brought Kelvin public acclaim. Faraday had studied the possible data rates for telegraphy depending on a cable's construction. The Cambridge mathematician and physicist George Gabriel Stokes contacted Kelvin in 1854 to ask his opinions on this. Kelvin made a detailed assessment of possible data rates, as well as the potential for a transatlantic cable's financial returns. He later joined the board of directors of the Atlantic Telegraph Company, improved the design of cables and travelled on ships to supervise the cable laying. The project had many setbacks and failures but achieved the first practical transatlantic telegraphy link in 1858.

One of Kelvin's other famous results was to calculate the age of the Earth. Reasoning that the laws of thermodynamics must have operated throughout history, he argued that the Earth had gradually cooled down from an early red-hot state. He calculated that it would have taken something like 20 to 400 million years to cool to its current temperature. Late in his life, he narrowed this figure to between 20 and 40 million years old. Following the discovery that radioactive decay releases heat, **Rutherford** pointed out that this would imply a much greater age for the Earth, which would have taken longer to cool due to the decay of radioactive elements in its interior.

In 1907, Boltwood demonstrated through radiometric dating that even some surface rocks on the Earth are at least 570 million years old. However, Kelvin never publicly accepted he was wrong about the Earth's age, arguing that he had a strong argument that the Sun was no more than 20 million years old. Like Helmholtz, he arrived at this vastly underestimated figure by assuming that the Sun's energy arose from gravitational collapse. The Sun's true energy source is now known to be nuclear fusion and modern estimates suggest it formed shortly before the Earth about 4.6 billion years ago.

Among Kelvin's many innovations, he invented a tide-predicting machine and an adjustable compass that could be calibrated to avoid errors arising from the increasing use of iron in ships. He was knighted in 1866 and he is buried close to **Newton** in Westminster Abbey.

(GEORG FRIEDRICH) BERNHARD RIEMANN
17 September 1826 – 20 July 1866

German mathematician whose work enabled Einstein to develop his general theory of relativity

'One would, of course, like to have a rigorous proof of this, but I have put aside the search for such a proof after some fleeting vain attempts because it is not necessary for the immediate objective of my investigation'

Riemann's comment in an 1859 lecture introducing the Riemann Hypothesis
(recorded in *Prime Obsession* by John Derbyshire, 2003)

Georg Riemann showed exceptional mathematical talent as a child. In 1846, he enrolled at the University of Göttingen to study theology (his father was a Lutheran minister), but he also attended maths lectures by **Gauss**, which convinced him to switch courses. He studied maths at the University of Berlin from 1847 then returned to Göttingen, where he earned his doctorate under Gauss in 1851 and in 1859 was appointed head of the maths department.

Riemann introduced that idea that physical reality might require a description with more than three space dimensions. In the mid-1850s, he introduced the many-dimensional geometry named after him. This turned out to be a vital tool for formulating **Einstein**'s general theory of relativity.

He also made some important contributions to number theory. He introduced a new function called the Riemann zeta function and showed that it has important implications for understanding the distribution of prime numbers. The zeta function has also been useful for problem solving in physics, probability theory and applied statistics. His 'Riemann hypothesis' is a conjecture regarding the zeros that occur in the zeta function – a rigorous proof of the conjecture eluded him, and proving it remains one of the most important unresolved problems in pure mathematics to this day.

During the last few years of his life, Riemann spent much of his time in Italy, where he died of tuberculosis aged only 39.

JOHN SNOW
15 MARCH 1813 – 16 JUNE 1858

English physician who promoted use of anaesthetics and pioneered epidemiology

'I found that nearly all the deaths had taken place within a short distance of the pump. There were only ten deaths in houses situated decidedly nearer to another street pump. In five of these cases the families of the deceased persons informed me that they always sent to the pump in Broad Street, as they preferred the water to that of the pumps which were nearer.'

Snow explains how he identified a water pump as the source of a cholera outbreak
(*Medical Times and Gazette*, 23 September 1854)

Hailing from York, Snow was an apprentice to a surgeon in Newcastle-upon-Tyne from the age of 14. In 1836, be began studying at the Hunterian School of Medicine in London and from 1837, he worked at the Westminster Hospital. He graduated with an MD from the University of London in 1844.

Snow pioneered the use of ether and chloroform as surgical anaesthetics, and was one of the first physicians to calculate suitable dosages. He personally administered chloroform to Queen Victoria during childbirth. But he is most famous for pinpointing the source of a cholera outbreak. At the time, before the introduction of the germ theory of disease, it was commonly believed that pollution or 'bad air' caused diseases such as cholera and the plague.

When a cholera outbreak occurred in London's Soho district in 1854, Snow interviewed local residents and showed that the victims had probably all drunk water from the same pump in Broad Street (now Broadwick Street). His analysis was persuasive enough to make the local council disable the pump by removing its handle. Snow died aged 45 following a stroke and is today commemorated with a memorial on Broadwick Street.

CLAUDE BERNARD
12 July 1813 – 10 February 1878

*French physiologist who is often considered the father of
modern experimental medicine*

'The physiologist is no ordinary man. He is a learned man, a man
possessed and absorbed by a scientific idea. He does not hear the
animals' cries of pain. He is blind to the blood that flows. He sees
nothing but his idea, and organisms which conceal from him the
secrets he is resolved to discover.'

From Bernard's *Introduction to the Study of Experimental Medicine* (1865)

Bernard began his career in Lyon where he held ambitions
of becoming a successful playwright. A literary critic in Paris advised
him against this, however and suggested he study medicine instead. He
followed the advice and after his studies he was appointed as an assistant
to the physiologist François Magendie at the Collège de France from 1841.
He succeeded Magendie as professor of medicine there in 1855.

Bernard's early work clarified the role in digestion of the pancreas,
which secretes juices that break down carbohydrates, proteins and fats. He
also showed that the main digestive processes occur in the small intestine
rather than the stomach, as previously assumed. In 1856, he discovered
glycogen, a carbohydrate molecule made primarily by the liver and the
muscles for long-term energy storage. He recognized that digestion
doesn't just break down big molecules – it also involves the build-up of
small molecules like sugars into complex ones like glycogen.

He also observed how the nervous system regulates blood flow and
body temperature by dilating and constricting blood vessels, and he
investigated the physiological effects of various drugs and poisons.
Several accounts, including the quote above, suggest that his vivisection
experiments were brutal. Bernard was also renowned for his scientific
rigour, urging scientists to ensure objectivity.

WILLIAM HENRY PERKIN
12 MARCH 1838 – 14 JULY 1907

English chemist who pioneered the invention and commercialization of synthetic dyes

'The aniline colours, which delight the eye, have attained much more importance to science. Their rays are the torch which enlightens the path of the explorer in the dark regions of the interior of the molecule, and the man who lit the torch was William Henry Perkin.'

From a lecture by German chemist Adolf von Baeyer (Nuremberg, early 1900s)

Perkin was born in London and studied at the Royal College of Chemistry in London (now part of Imperial College) from the age of only 15. He became an assistant to the chemist August Wilhelm von Hofmann who assigned him the task of synthesizing quinine, in high demand for the treatment of malaria, from organic compounds such as aniline ($C_6H_5NH_2$).

In 1856, working at a basic home laboratory, Perkin discovered that by oxidizing aniline, he could produce a dark precipitate. By rinsing the container with a solvent, he created a substance with an intense purple colour. This turned out to be an effective permanent dye for silk. Perkin applied for a patent for the dye (mauvine) and backed by his father and brother, set up a dye works in London to manufacture it commercially, marking the first time a synthetic dye was manufactured in a purpose-built chemical plant anywhere in the world.

Perkin later invented and marketed other synthetic dyes. In 1869, he discovered a method to cheaply create the brilliant red dye alizarin from anthracene; previously, this dye had been extracted from madder crops. He sold his dye factory when he was 37, already a wealthy man, and devoted the rest of his life to basic chemical research.

(FRIEDRICH) AUGUST KEKULÉ VON STRADONITZ
7 September 1829 – 13 July 1896

*German chemist who founded the science of
structural chemistry*

'I turned my chair to the fire and dozed. Again the
atoms were gambolling before my eyes ... My mental
eye, rendered more acute by the repeated visions of the
kind, could now distinguish larger structures of manifold
confirmation: long rows, sometimes more closely fitted
together all twining and twisting in snake like motion.
But look! What was that? One of the snakes had seized
hold of its own tail, and the form whirled mockingly
before my eyes.'

Kekulé describes how he discovered the structure of benzene
(*Serendipity: Accidental Discoveries in Science* by Royston M. Roberts, 1989)

Kekulé attended the University of Giessen planning to
become an architect, but he was inspired to change his course to chemistry
after hearing lectures by **Liebig**. In 1856, he took a teaching post in
Heidelberg, where he also set up a home laboratory, and in 1858 he was
appointed professor at the University of Ghent.

Kekulé was first to recognize that molecules consist of atoms that
are linked together according to the 'combining power' (valency) of each
atom. He proposed that carbon has a valency of four, meaning that it can
combine with four other atoms, and that carbon itself can bond with up to
three of these four bonds of a carbon atom, so that long chains of carbon
atoms could form.

Allowing hydrogen one bond, oxygen two and nitrogen three, Kekulé
finally made sense of many organic compounds and their various isomers
(molecules with the same atoms arranged differently) in terms of structural
formulae. Famously, the ring structure of benzene (C_6H_6) came to him in
a dream as he dozed by the fire.

RUDOLF LUDWIG KARL VIRCHOW
13 October 1821 – 5 September 1902

German doctor, biologist and politician regarded as the founder of modern pathology

'The hero who for the past twenty years has held undisputed pre-eminence in the realm of science is now translated to the Valhalla of his peers. His was the last great figure remaining to us of those who carried the torch of honest inquiry into the dark places of traditional dogma and mediaeval superstition. The universal reverence of mankind was his reward.'

From Virchow's obituary (*British Medical Journal*, 13 September 1902)

Virchow was born in Schivelbein, Prussia, which is now in Poland. He studied medicine in Berlin and after holding a junior post at Berlin's Charité Hospital, he was appointed professor of pathological anatomy at Würzburg from 1849 until 1956, when he became professor of pathology at the University of Berlin.

Virchow made important contributions to the understanding of a host of diseases, for instance, by helping establish leukaemia as a disease linked to white blood cells. His medical knowledge and prolific use of the microscope vastly improved public health. As a city councillor in Berlin from 1859, he worked to improve hygiene by building new hospitals and redesigning the sewer systems. He was also interested in archaeology and anthropology, and was influential in opposing the concept that there is a pure German race.

His greatest legacy however, was the identification of the cell as the fundamental unit of life. He took the view that disease originates in cells, or at least is a consequence of cells experiencing abnormal circumstances. However, Virchow did not accept **Darwin**'s theories, arguing that there was a lack of fossil evidence for a common ancestor of man and apes.

WILHELM MAXIMILIAN WUNDT
16 August 1832 – 31 August 1920

German physician often considered a founder of modern experimental psychology

'Others maintain that there are no laws at all in the mental sphere. They say that the essential difference between natural and mental sciences consists in the fact that only the former can be reduced to definite laws ... This opinion rests obviously on a mistaken use of the conception of law. We are only allowed to consider those regularities in phenomena as according to law, which always repeat themselves in exactly the same manner. But there are in reality no such laws, not even in the natural sciences.'

From Wundt's *An Introduction to Psychology* (1912)

Born in Neckarau, near Mannheim, Wundt studied medicine at the universities of Tübingen, Heidelberg and Berlin from 1851 to 1856. He then joined the faculty at Heidelberg, where he became an assistant to **Helmholtz** and later assistant professor. In 1875 he moved to the University of Leipzig, where in 1879, he founded the first laboratory devoted entirely to experimental psychology.

Before Wundt, it was commonly accepted that psychology cannot be an exact science, nor even an experimental one, because thought and behaviour are not mathematically expressible and different thoughts can't be isolated. Wundt argued that it is possible to study consciousness aided by the experimental protocols of the natural sciences and, by founding his experimental laboratory, he established psychology as a science in its own right. His early experiments involved measuring variations in reaction times among different people, and he also explored the nature of religious belief and abnormal behaviours.

A prolific writer, Wundt's works amounted to around 53,000 pages and covered a vast range of subjects including physiology, hypnotism, vision, spiritualism, the effects of poisons, history, politics, ethics and religion. He also described the Wundt illusion, an optical illusion in which two straight parallel lines look like they're bowed inwards when superimposed on a background of bent lines.

ERNST HEINRICH PHILIPP AUGUST HAECKEL
(OR VON HAECKEL)
16 February 1834 – 9 August 1919

German naturalist and philosopher who mapped relationships between all life forms and promoted Darwin's evolutionary theories

'In a word, ecology is the study of all those complex interrelations referred to by Darwin as the struggle for existence'

From Haeckel's *Morphology of Organisms* (1866)

Haeckel was born in Potsdam, then in Prussia. He studied medicine in Berlin and Würzburg, earning a doctorate in medicine in 1857, but then changed his plans and studied zoology at the University of Jena, where he later became professor of anatomy from 1862 to 1909.

From 1859 onwards, Haeckel made many expeditions to the Canary Islands, Turkey, Greece, Egypt and Norway, and he also met **Darwin**, **Huxley** and **Lyell**. He became a prolific writer; in Germany before the First World War, more people learned of evolutionary theory from Haeckel's writings than from any other source. However, Haeckel did not support the idea of natural selection – his own views were closer to those of **Lamarck**, who advocated the idea that an organism can pass on to its offspring the characteristics it acquired during its lifetime.

A talented illustrator, Haeckel discovered, described and named thousands of new species during his lifetime, and made beautiful drawings of them, particularly of radiolarians (marine protozoa). He also coined the term 'ecology'. He was prone to jump to conclusions, however, describing in great detail what undiscovered remains of ancient humans might look like. He also promoted the popular misconception that an animal embryo replays all the earlier evolutionary stages of the creature, and his statement that 'politics is applied biology', as well as his flawed views on human evolution, were later used by the Nazis to justify racism and social Darwinism.

CHARLES ROBERT DARWIN
12 February 1809 – 19 April 1882

English naturalist who developed the theory of evolution by natural selection

'When on board HMS *Beagle*, as naturalist, I was much struck
with certain facts in the distribution of the inhabitants of South
America, and in the geological relations of the present to the past
inhabitants of that continent. These facts seemed to me to throw
some light on the origin of species – that mystery of mysteries,
as it has been called by one of our greatest philosophers ...
I am fully convinced that species are not immutable; but that
those belonging to what are called the same genera are lineal
descendants of some other and generally extinct species ...
Furthermore, I am convinced that Natural Selection has been
the main but not exclusive means of modification.'

From the introduction to Darwin's *On the Origin of Species* (1859)

Darwin came from a wealthy and
well-connected family based in Shrewsbury,
Shropshire. His maternal grandfather was the china
manufacturer Josiah Wedgwood, while his paternal
grandfather was Erasmus Darwin, one of the leading
intellectuals of 18th century England. Charles Darwin
studied medicine at the University of Edinburgh but
later changed his plans and moved to Cambridge
University to study divinity.

He also developed a strong interest in natural history, encouraged by
Cambridge botany professor John Henslow. On Henslow's recommendation, Darwin was appointed as naturalist on the HMS *Beagle*, which set
sail in 1831 to survey the coast of South America and the Pacific. During
the journey, he read **Lyell**'s *Principles of Geology* published in the previous
year. The book persuasively argued the case for **Hutton**'s uniformitarianism
concept, which held that the geological forces shaping the Earth today are the
same ones that have always operated. That implied that the Earth could not
be a few thousand years old, as Biblical accounts suggested, but must be at
least millions of years old.

In South America, Darwin was struck by the geological formations whose history he saw 'through Lyell's eyes'. He was also intrigued by interesting patterns in the way species gradually change from region to region, especially on the Galápagos Islands, 970 km (600 miles) west of Ecuador. For example, he noted the islands hosted more than a dozen species of finch, each thriving in a different part of the islands. Darwin thought it extremely unlikely that all the species had appeared independently; he was convinced that they must have evolved from a parent species from the mainland.

On his return to England in 1836, Darwin tried to explain how this evolution might have occurred. Part of his inspiration came from the ideas of Thomas Malthus, the English scholar who argued that human populations increase or decrease in response to various environmental factors such as famine. Darwin proposed that in an analogous way, animals or plants that are best suited to their environment are more likely to survive and reproduce, passing on their favourable characteristics to their offspring so that gradually, a species changes over time.

Darwin continued to mull over this 'natural selection' theory for two decades and after hearing that **Wallace** had come up with the same basic concept, they both arranged for scientific papers to be presented to a meeting of the Linnean Society in London in 1858. The following year, Darwin published his famous book about natural selection. Its full title was *On the Origin of Species by Means of Natural Selection, or the Preservation of Favoured Races in the Struggle for Life*, although this was shortened to *The Origin of Species* for the sixth edition of 1872.

The book was popular but controversial. Although Darwin had shied away from discussing the implications of his theory for human origins, a logical extension was that people evolved from other animals such as apes. Cartoonists published absurd parodies of the theory showing Darwin's head on an ape's body. But Darwin's concept of evolutionary adaptation through natural selection later became widely accepted.

Even the most unanticipated discoveries in genetics and the fossil record have since supported or extended Darwin's central ideas – all life is related, species change over time in response to natural selection and new forms replace those that came before. Darwin died on 19 April 1882 and is buried in Westminster Abbey.

ALFRED RUSSEL WALLACE
8 January 1823 – 7 November 1913

British naturalist who developed a theory of evolution independently of Darwin

'The powerful retractile talons of the falcon- and the cat-tribes have not been produced or increased by the volition of those animals; but among the different varieties which occurred in the earlier and less highly organized forms of these groups, those always survived longest which had the greatest facilities for seizing their prey ... There is a tendency in nature to the continued progression of certain classes of varieties further and further from the original type.'

From Wallace's *On the Tendency of Varieties to Depart Indefinitely From the Original Type* (1858)

Born in the village of Llanbadoc, Monmouthshire, Wallace worked as a surveyor in England and Wales before being hired as a master at the Collegiate School in Leicester in 1844. That year he became friends with a local naturalist, Henry Walter Bates, and four years later, they departed on an expedition to South America having read accounts of **Darwin**'s voyage on the *Beagle* and **Humboldt**'s travels there.

In South America, Wallace gathered thousands of unique animal specimens, mostly birds, beetles and butterflies. He lost most of them on the return journey when his ship sank, the passengers rescued by a passing boat. Undeterred, in 1854 Wallace embarked on a new expedition to the Malay Archipelago (now Malaysia and Indonesia) where he accumulated around 125,660 specimens, including many new to science.

During this expedition, Wallace became convinced that species evolved because the fittest individuals survived and reproduced, passing advantageous characteristics to their offspring. He wrote about this idea to Darwin, who had been working on it for many years but had not published it. This prompted Darwin to publish his classic work *On the Origin of Species* in 1859.

FRANCIS GALTON
16 FEBRUARY 1822 – 17 JANUARY 1911

English polymath and explorer who coined the term 'eugenics'

'It is notorious that the same discovery is frequently made
simultaneously and quite independently, by different persons.
Thus, to speak of only a few cases in late years, the discoveries
of photography, of electric telegraphy, and of the planet Neptune
through theoretical calculations, have all their rival claimants.
It would seem, that discoveries are usually made when the time
is ripe for them – that is to say, when the ideas from which they
naturally flow are fermenting in the minds of many men.'

From Galton's book *Hereditary Genius* (1869)

A cousin of Darwin, Galton was born in Birmingham and studied
medicine there and at King's College, London. He later attended Cambridge
University to study maths. He abandoned further studies after receiving
an inheritance when his father died, his independent wealth allowing him
to travel widely in Africa and the Middle East. His descriptions of their
geography won him wide acclaim.

Galton contributed to diverse fields of science. In meteorology,
he identified and named anticyclones, the large-scale circulation of
winds around a central region of high atmospheric pressure. He also
introduced the first popular weather maps to newspapers. Inspired by
Darwin's publication of *On the Origin of Species*, Galton began studies
of the variations of traits in people, something that Darwin had barely
touched on, and to analyse these using new statistical techniques. Using
biographical sources, he tested whether eminent men have more eminent
relatives than average and concluded that they did – that suggested to him
that ability is an inheritable trait.

In 1883, Galton coined the term 'eugenics' to describe efforts to improve
the genetic heritage of human populations. He advocated a point-scoring
system to identify high-quality families and recommended introducing
incentives for members of high-ranking families to marry young and have
children. Galton was also very active in the running of scientific societies,
including the British Association for the Advancement of Science, and he
was knighted in 1909.

ROBERT WILHELM EBERHARD BUNSEN
30/31 March 1811 – 16 August 1899

German chemist who identified new elements and pioneered spectroscopy

'At present Kirchhoff and I are engaged in a common work which doesn't let us sleep ... Kirchhoff has made a wonderful, entirely unexpected discovery in finding the cause of the dark lines in the solar spectrum ... Thus a means has been found to determine the composition of the Sun and fixed stars with the same accuracy as we determine sulphuric acid, chlorine, etc, with our chemical reagents.'

From a letter Bunsen wrote on 15 November 1859
(*Discovery of the Elements* by Mary Elvira Weeks, 1933)

Göttingen-born Bunsen studied science and maths at the University of Göttingen, where he earned a PhD in 1831. After travelling in Europe, he became a lecturer at Göttingen, and in 1836 he succeeded **Wöhler** at the Polytechnic School of Kassel. He was later professor at Marburg and Heidelberg.

In early work, Bunsen used electrolysis to produce pure samples of metals, such as chromium, sodium and lithium. He also prepared toxic, flammable organic compounds containing the cacodyl group $(CH_3)_2As$. During this work, he suffered from arsenic poisoning and an explosion in his lab made him lose sight in one eye.

Much of Bunsen's most significant work was done in collaboration with **Kirchhoff**, who suggested in 1859 that they used a prism instrument to split light from heated elements into its constituent colours. Although studies of these characteristic colours had been done before, they were not rigorous and systematic. Within months, the pair had built their instrument, a very basic kind of spectroscope, and they used it to measure the spectra of many elements including sodium and potassium.

Bunsen noticed that highly pure samples of the elements gave unique spectra that would always allow their identification. During his experiments, he observed that samples of mineral water gave off blue spectral emission lines with wavelengths he hadn't seen before, and he suspected it signalled a previously undiscovered element. To isolate it, he distilled around 40 tonnes of mineral water and, in 1860, isolated 17 grams (two thirds of an ounce) of the new element. He named the element caesium after the Latin word 'caesius', deep blue.

The following year he discovered rubidium, again through spectroscopy. Bunsen and Kirchhoff realized that this technique could identify the composition of the Sun. By simultaneously observing the Sun's spectrum and flame spectra of elements in the lab, they showed that the colours of bright emission lines in heated gases match the dark bands of absorption when white light passes through the same cool gas. That showed it would be possible to identify the elements responsible for dark bands already noted in the solar spectrum. Kirchhoff then identified many chemicals in the Sun.

To most people, Bunsen is remembered best for his invention of the Bunsen burner, a mainstay of school chemistry laboratories. In fact, he and his assistant Peter Desaga only made improvements to existing lab burners to create a hotter, cleaner flame.

In personality, Bunsen was renowned as an excellent teacher and amiable colleague who always avoided conflict. After meeting Bunsen for the first time, **Fischer**'s wife Agnes supposedly commented: 'First, I would like to wash Bunsen, and then I would like to kiss him because he is such a charming man'.

LOUIS PASTEUR
27 December 1822 – 28 September 1895

French chemist and microbiologist who developed the germ theory of disease

'Science knows no country because knowledge belongs to humanity, and is the torch which illuminates the world. Science is the highest personification of the nation because that nation will remain the first which carries the furthest the works of thought and intelligence ... One does not ask of one who suffers: What is your country and what is your religion? One merely says: You suffer, this is enough for me: you belong to me and I shall help you.'

Pasteur quoted in *Louis Pasteur: Free Lance of Science* by Rene J. Dubos (1950)

Pasteur earned a doctorate at the École Normale Superiéure in Paris in 1847 for studies in the fledgling science of crystallography. He then held to teaching and research posts in Dijon and Strasbourg, and he became professor of chemistry at the University of Lille in 1854. He was appointed director of scientific studies at the École Normale in 1857, and later moved to the Sorbonne as professor of chemistry from 1867. During the last years of his career, he directed Paris's Pasteur Institute founded to develop treatments for disease and named in his honour.

In his earliest work, Pasteur made profound contributions to structural chemistry. He discovered molecular asymmetry in studies of tartaric and racemic acid, which were already known to have the same chemical formulae. But their crystalline salts, in solution, had very different optical properties. One rotated a beam of polarized light to the right, while the other had no effect on polarized light. Pasteur correctly concluded that racemic acid was a mixture of two mirror image molecules, whose optical effects cancelled out, while tartaric acid contained just one of these molecules, creating an asymmetric twisting effect on light.

By 1856, Pasteur had begun work on fermentation. He showed that microorganisms cause fermentation in wine, beer and milk so that fermentation is not, as **Liebig** proposed, a purely chemical effect. His experiments to determine the source of the microbes proved that they were introduced from the environment, which went against popular contemporary notions that they arose by spontaneous generation. Pasteur also showed that the bacteria could be removed by boiling and then cooling a liquid, a sterilization process now called pasteurization.

In 1865, Pasteur was tasked with investigating an epidemic of disease in silkworms, which was devastating the silk industry of southern France. He successfully identified a parasitic infection as the cause, and suggested a solution (selecting disease-free eggs) that restored the industry to health.

But he is best known for establishing the germ theory of disease, which holds that microorganisms from the environment can invade the body to cause infectious disease, as well as his work on vaccinations. In the 1870s, he tried to solve the serious problem of chicken cholera, which could spread through an entire chicken farm in just a few days. He and his colleagues cultured chicken cholera bacteria and left them standing in hot weather. When they later injected them into chickens, they didn't develop the disease when injected with a larger amount of live microbes. Pasteur had accidentally discovered the concept of attenuated vaccines that confer immunity.

In subsequent work on anthrax, a disease that particularly affected sheep and cattle, Pasteur successfully demonstrated a vaccine in 1881. One of his most famous innovations was a vaccine for rabies, which he produced by growing rabies viruses in rabbits then weakening them by drying the affected nerve tissue. He tested the vaccine on dogs before successful human trials commenced from 1885 onwards. This was only the second vaccine effective against a serious human disease to be invented since **Jenner** had developed the smallpox vaccine many decades earlier.

While Jenner's vaccine had relied on the discovery of a natural weaker form of the disease agent (cowpox), Pasteur showed that by artificially altering the potentially fatal infectious microbe itself, it was possible to confer immunity. His germ theory led to dramatic improvements in public health by inspiring **Lister** to promote antiseptic surgery. Today, Pasteur is often regarded as the father of the germ theory of disease as well as the science of bacteriology, together with **Koch**.

THOMAS HENRY HUXLEY
4 MAY 1825 – 29 JUNE 1895

*English biologist who promoted Charles Darwin's
theory of evolution*

'If the question is put to me would I rather have a miserable
ape for a grandfather or a man highly endowed by nature and
possessed of great means of influence and yet who employs
these faculties and that influence for the mere purpose of
introducing ridicule into a grave scientific discussion,
I unhesitatingly affirm my preference for the ape'

Huxley's response to English bishop Samuel Wilberforce when the
bishop sarcastically asked whether he had descended from an ape on
his grandmother's side or his grandfather's
(*Life and Letters of Thomas Henry Huxley*, 1900)

Huxley received little formal education during his early
years but taught himself science, history and philosophy. Aged 15, he began
a medical apprenticeship that led to a scholarship to study at Charing Cross
Hospital and from 1846 to 1850, he was assistant surgeon on a Royal Navy
mission to map the seas around Australia and New Guinea. From 1854
until 1885, he was professor at the School of Mines in London.

On his South Seas expedition, Huxley collected and studied marine
animals, and this work led him into the circles of eminent scientists
including **Lyell** and **Darwin**. He became an outspoken advocate for
Darwin's evolutionary theories, which earned him the nickname of
'Darwin's bulldog', and he took part in a celebrated debate on the issue
with bishop Samuel Wilberforce in Oxford in 1860.

However, Huxley was critical of some aspects of Darwin's theory,
suggesting evolution included rapid 'jumps' as well as slow, continuous
change. His best-known written work, *Evidence as to Man's Place in Nature*
(1863), applied evolutionary theory to humans, something Darwin had
side-stepped.

(PIERRE) PAUL BROCA
28 June 1824 – 9 July 1880

French physician and anatomist who discovered the brain's speech production centre

'He could no longer produce but a single syllable, which he usually repeated twice in succession; regardless of the question asked him, he always responded: tan, tan, combined with varied expressive gestures. This is why, throughout the hospital, he is known only by the name Tan.'

Broca's description of one patient with brain damage (around 1861)

Broca was born in Sainte-Foy-la-Grande in Gironde, southwest France. He studied medicine at the University of Paris then specialized in surgery. Around 1861, he studied a patient with neurological problems and an inability to say any words except 'tan'. After the patient died, Broca showed in a post-mortem examination that the patient had a lesion on the left frontal lobe of his brain (now named 'Broca's area'), suggesting this is the brain's speech production centre.

This was one of the first conclusive demonstrations that localized regions of the cerebral cortex control different bodily functions. Broca was also a pioneer of physical anthropology. He founded the Anthropology Society of Paris in 1859 and the Parisian School of Anthropology in 1876, helping to establish the subject as a rigorous branch of science. An early supporter of **Darwin**'s still highly controversial theory of evolution, Broca made precise measurements of skulls discovered at prehistoric sites and argued that they represented primitive ancestors that later evolved into modern man.

He also researched a wide range of different subjects including cancer pathology, infant mortality and treatments for brain aneurysms. He himself died of an aneurysm and his brain is still preserved today at the Museum of Man in Paris.

GREGOR JOHANN MENDEL
20 JULY 1822 – 6 JANUARY 1884

*Austrian scholar and Augustinian priest who
determined laws of heredity*

'Experiments on ornamental plants undertaken in previous years
had proven that, as a rule, hybrids do not represent the form exactly
intermediate between the parental strains. Although the intermediate
form of some of the more striking traits, such as those relating to
shape and size of leaves, pubescence of individual parts, and so forth,
is indeed nearly always seen, in other cases one of the two parental
traits is so preponderant that it is difficult, or quite impossible, to
detect the other in the hybrid ... Those traits that pass into hybrid
association entirely or almost entirely unchanged, thus themselves
representing the traits of the hybrid, are termed *dominating*, and
those that become latent in the association, *recessive*.'

Mendel outlines his discoveries in plant breeding
(*Gregor Mendel's Experiments on Plant Hybrids* by Alain Corcos
and Floyd Monaghan, 1993)

In 1843, Mendel joined the Augustinian
Order at St Thomas's Abbey in Brünn (now Brno in
the Czech Republic) and began theological studies.
He was ordained as a priest in 1847, taught himself
natural sciences and later studied sciences and
maths at the University of Vienna. After returning to
Brünn, he began the experiments that would make
him world famous after his death.

Mendel selectively cross-bred common peas (*Pisum sativum*) over many
generations to study the patterns of inheritance of seven traits including
flower colour, stem length and seed shape. Before Mendel's work, heredity
was regarded as a mixing and dilution of the different parental traits. But
Mendel showed this is not true. For instance, the flower colours of pea
plants in successive generations were always purple or white, not some
intermediate between the two.

Mendel recognized that each trait was determined by two factors of
inheritance. Each parent has a pair of factors and passes one of these to
their offspring. By measuring how often certain traits appeared, Mendel

showed that they were inherited according to chance, and that the traits were all inherited independently. All the traits had a dominant form that masked the alternative one, but either could appear in future generations.

For example, he noticed that when he crossed pea plants with long stems with dwarf plants, all the offspring plants were tall. When these hybrids were crossed with each other, the tall to short ratio of the offspring was 3:1. Mendel's ingenious insight was that this is due to 'dominant' and 'recessive' factors, which today are known to be dominant and recessive genes. His results on heredity are summarized in 'Mendel's laws', the law of segregation and the law of independent assortment.

In 1865, Mendel presented his findings to the Society for the Study of the Natural Sciences in Brünn. The following year, a paper on his work was widely circulated to scientific associations, libraries and academics. Despite this, his work was largely ignored until around 1900, when other scientists replicated the findings independently. Around that time, Mendel's work was rediscovered and acknowledged.

Mendel's results resolved a worrying problem in Darwin's theory of evolution. If successive generations of animals or plants had merely blended traits from their parents, it was difficult to understand how natural selection could exert its effect. The discovery that traits do not blend, but remain distinct, meant that natural selection could indeed work slowly and effectively on natural variation. Darwin died in 1882 without knowing this weakness in his theory had been resolved. In the 1930s and 1940s, the combination of Mendelian genetics with Darwin's theory of natural selection resulted in the modern synthesis of evolutionary biology.

One aspect of Mendel's work has remained controversial, however. Statistically, it is unlikely that the heredity patterns he observed were as precise as he reported. While his conclusions about the laws of heredity were correct, he may have noticed the approximate relation early on and then biased subsequent results to conform to it.

After publishing his work on plants, Mendel carried out research in meteorology, horticulture and astronomy. He also attempted to extend his plant work to animals by creating hybrid bees, although he failed to find clear patterns of inheritance. In 1868, Mendel was elected abbot of St Thomas's Abbey and largely stopped his scientific work due to the demands of civic responsibilities and a protracted dispute with the civil government over taxes on religious institutions.

JOSEPH LISTER
5 April 1827 – 10 February 1912

British surgeon who pioneered antiseptic surgery

'When it has been shown by the researches of Pasteur that the
septic property of the atmosphere depended not on the oxygen,
or any gaseous constituent, but on minute organisms suspended
in it, which owed their energy to their vitality, it occurred to me
that decomposition in the injured part might be avoided without
excluding the air, by applying as a dressing some material capable
of destroying the life of the floating particles'

From a report by Lister in the *British Medical Journal* (1867)

Lister studied medicine at the University
of London, graduating in 1852. He practised as
a surgeon in Edinburgh before being appointed
professor of surgery at Glasgow Royal Infirmary in
1860. In 1869, he returned to Edinburgh and was
made professor of surgery at King's College, London,
in 1877.

Following the introduction of anaesthetics
around the 1840s, surgical operations were more frequently performed
but many patients suffered fatal infections afterwards. Doctors believed
something in the air caused the infections, so they dressed wounds tightly
or irrigated them with water to keep the air out. After reading about
Pasteur's germ theory in 1864, however, Lister realized that post-operative
infection was probably the result of germs entering the wound.

Hearing that carbolic acid treatment of sewage in Carlisle had effectively
reduced infectious disease in the local community, Lister showed in 1865
that the same chemical successfully reduced infection when applied to
wounds. It soon became clear that surface germs were the key problem,
not airborne ones. He went on to introduce surgical use of sterile catgut
and silk, as well as gauze dressings, and antisepsis became a driving force
for more ambitious, safer surgery.

HEINRICH SCHLIEMANN
6 January 1822 – 26 December 1890

German amateur archaeologist who excavated Troy

'In excavating this wall further and directly by the side
of the palace of King Priam, I came upon a large copper
article of the most remarkable form, which attracted
my attention all the more as I thought I saw gold
behind it ... While the men were eating and resting,
I cut out the treasure with a large knife.'

From Schliemann's report of his discoveries at the Turkish site of Hisarlik (May 1873)

Schliemann had a classic rags-to-riches life story. Born in Neubukow, Mecklenburg-Schwerin, he took a job as a cabin boy on a steamer bound for Venezuela in 1841, and when the ship sank in rough weather off the coast of the Netherlands, he made it to the shore and developed a successful business career in Amsterdam and later St Petersburg. He then moved to California, where he amassed a fortune speculating on stock markets.

After returning to Russia and further business success, partly by supplying commodities to the Russian government during the Crimean War, Schliemann decided to pursue a childhood ambition to discover the site of Troy, the setting for the Trojan War described by Homer. Hearing from English amateur archaeologist Frank Calvert that the remains of Troy might be at a site called Hisarlik on Calvert's family's property in western Turkey, Schliemann began excavating the site in 1871.

In 1873, he found a cache of gold and other artefacts that he believed were the treasures of Homeric king Priam. While Schliemann's work added credence to the idea that Homer's *Iliad* described some real historical events, his rough excavations probably destroyed much of the interesting information preserved at Hisarlik.

ALFRED BERNHARD NOBEL
21 October 1833 – 10 December 1896

Swedish chemist who invented dynamite and financed the Nobel prizes

'The whole of my remaining realizable estate shall be dealt with in the following way: the capital, invested in safe securities by my executors, shall constitute a fund, the interest on which shall be annually distributed in the form of prizes to those who, during the preceding year, shall have conferred the greatest benefit on mankind ... It is my express wish that in awarding the prizes no consideration be given to the nationality of the candidates, but that the most worthy shall receive the prize, whether he be Scandinavian or not.'

Excerpt from the will of Alfred Nobel

Nobel was born in Stockholm and his family moved to St Petersburg, Russia, when he was still a young child. His father ran an engineering and construction business that supplied equipment to the Russian army. Alfred Nobel was privately tutored in natural sciences and literature, and he was proficient in five languages by the age of 17.

Nobel went on to study chemistry in France and the US. While living in Paris, he met the Italian chemist, Ascanio Sobrero, who had invented nitroglycerine, and Nobel was intrigued to find out if he could develop nitroglycerine as a commercially and technically useful explosive. In 1856, at the end of the Crimean War, his father's business floundered when military orders dried up and most of Nobel's family decided to return to Sweden in 1863.

Back in Stockholm, Nobel pursued developing nitroglycerine as an explosive and was granted the first patent for its use as an industrial explosive (blasting oil) in 1863. But the Swedish government banned his experiments within the city following accidents that killed several people, including Nobel's own younger brother, Emil.

Nobel continued experiments on nitroglycerine on a lake barge, and in 1864 he began mass production of nitroglycerine. He also discovered in experiments that if he mixed liquid nitroglycerine with fine sand called kieselguhr, it was possible to develop a paste that could be shaped into rods to place in drill holes. This made it easier and safer to handle. He patented this invention – which he named 'dynamite' – in 1867. Dynamite went on to dramatically reduce the cost of construction projects such as tunnel boring and road building.

In other experiments he formulated another nitrate explosive, gelignite, a jelly-like substance that turned out to create more powerful blasts than dynamite. He patented gelignite in 1876. The success of his explosives manufacturing led Nobel to build dozens of factories across Europe and in the US. He also experimented with creating synthetic rubber and leather, as well as artificial silk. By the time of his death in 1896 he had been granted 355 patents.

Reports suggest that Nobel was concerned about negative press. In 1888, when his brother Ludvig died, the staff of a French newspaper thought Alfred himself had died and published an obituary stating that he 'became rich by finding ways to kill more people faster than ever before'. Perhaps in response to this, Nobel signed a will in 1895 stating that the majority of his vast fortune should be used after his death to establish the Nobel prizes, which would reward some of the world's most ground-breaking achievements that benefit mankind.

His will decreed that his money should be invested and the annual interest paid to winners in five fields: 'The said interest shall be divided into five equal parts, which shall be apportioned as follows: one part to the person who shall have made the most important discovery or invention within the field of physics; one part to the person who shall have made the most important chemical discovery or improvement; one part to the person who shall have made the most important discovery within the domain of physiology or medicine; one part to the person who shall have produced in the field of literature the most outstanding work in an ideal direction; and one part to the person who shall have done the most or the best work for fraternity between nations, for the abolition or reduction of standing armies and for the holding and promotion of peace congresses.'

The first Nobel prizes were awarded in 1901. The science prizes are widely acknowledged to have spurred research by rewarding ingenious, sometimes obscure research, often by small and dedicated teams that are not motivated by commercial benefits. The prize money, which depends on the Nobel Foundation's investment income, is typically nearly £1,000,000 (US$1,500,000). Each prize is often shared by up to three people.

DMITRI IVANOVICH MENDELEYEV
8 February 1834 – 2 February 1907

Russian chemist who established the periodic table of the elements

'I began to look about and write down the elements with their atomic weights and typical properties, analogous elements and like atomic weights on separate cards, and soon this convinced me that the properties of the elements are in periodic dependence upon their atomic weights; and although I had my doubts about some obscure points, yet I have never doubted the universality of this law, because it could not possibly be the result of chance.'

From Mendeleyev's *Principles of Chemistry* (1905, vol 2)

Mendeleyev was born into a large family of more than a dozen children near Tobolsk, Siberia, where his family suffered severe hardship. His father was a teacher until he went blind and lost his job, so to make ends meet his mother restarted her family's abandoned glass factory, although this was subsequently destroyed by fire. Around the same time, the father died and the family moved to St Petersburg, where Mendeleyev enrolled to study at the Main Pedagogical Institute. After spending time on the Black Sea coast to recover from tuberculosis, he studied in Heidelberg where he briefly collaborated with **Bunsen**.

Mendeleyev was appointed as professor at the St Petersburg Technical Institute in 1863 and then at the University of St Petersburg in 1866. He resigned his post at the university in 1890 when the authorities expressed fears that his liberal views were corrupting students, but he was later appointed as director of the Bureau of Weights and Measures, tasked with introducing new standards for vodka production.

Mendeleyev won world fame for proposing that the chemical elements form a periodic table, although he was not the first to do so. In 1863, a total of 56 chemical elements had been discovered and by this time, several scientists had noticed that the chemical elements display periodically repeating patterns in their chemical properties. Independently, Mendeleyev attempted to classify elements while he was writing his major work *The Principles of Chemistry*, which was first published in two volumes between 1868 and 1870.

While he was writing the book, he was shuffling papers on his desk with the known chemical elements and their properties noted down. He noticed that if he arranged them in a table of rows with atomic weight increasing from left to right, he could make the columns contain elements with similar chemical properties. Crucially, he recognized that the table contained gaps that must correspond to elements not yet discovered, and that their chemical properties could be predicted to some extent in advance. Over many years, he persistently refined his version of the periodic table to improve its consistency.

Mendeleyev predicted the properties of the elements now called gallium, scandium and germanium, and all three went on to be discovered by 1886. He did not clearly understand the reason why periodic relationships appear for chemical elements, something that would not become apparent until scientists discovered that atoms consist of a dense inner nucleus surrounded by electrons that gradually fill up a series of shells as the atomic mass increases. The chemical properties of an element depend on the number of electrons in the outermost shell.

In other work, Mendeleyev studied the expansion of liquids at increasingly higher temperatures and derived a formula similar to Gay-Lussac's law of volumes. During a solo balloon flight without any training, he photographed a solar eclipse in 1887. He was a driving force in establishing the metric system of units across the Russian Empire and he also studied the composition of petrol, concluding that hydrocarbon fuels only form deep in the Earth.

Throughout his career, Mendeleyev was committed to helping establish thriving industry in Russia and he helped set up the first oil refinery there. For his work on the periodic table, the Nobel Committee for Chemistry recommended that he be awarded the 1906 Nobel prize for chemistry, but after heated debate about his contributions, his award was blocked, partly due to pressure from **Arrhenius**. Some reports suggest that Arrhenius was embittered by Mendeleyev's criticisms of his scientific work and that this motivated his opposition to the award. The element mendelevium, with atomic number 101, was named in his honour.

JAMES CLERK MAXWELL
13 June 1831 – 5 November 1879

Scottish physicist and mathematician who formulated a ground-breaking theory of electromagnetism

'This characteristic of modern experiments – that they consist principally of measurements – is so prominent, that the opinion seems to have got abroad, that in a few years all the great physical constants will have been approximately estimated, and that the only occupation which will then be left to men of science will be to carry on these measurements to another place of decimals ... But we have no right to think thus of the unsearchable riches of creation, or of the untried fertility of those fresh minds into which these riches will continue to be poured.'

From a lecture by Maxwell in Cambridge, October 1871
(*The Scientific Papers of James Clerk Maxwell*, 2003)

James Clerk Maxwell showed an early talent for mathematics and studied maths, physics and moral philosophy at the University of Edinburgh. He then studied maths at Cambridge University, beginning in 1850. In 1856, he was appointed as professor at Marischal College in Aberdeen (later part of Aberdeen University). In 1871, he became the first Cavendish Professor of Experimental Physics at the University of Cambridge, where he set up the now famous Cavendish Laboratory.

As young as 18, Maxwell revived **Young**'s theory of colour vision and showed that the colours red, green and blue can combine to create white light. In astronomy, he suggested correctly that the rings of Saturn must be composed of small particles to remain in stable orbits. He also demonstrated that molecules in a gas don't travel with the same speed but have a bell curve of energies, and that this kinetic energy is the basis of heat (some scientists had argued that heat is a kind of fluid).

But Maxwell is best known for his mathematical theory of electromagnetism, which for this branch of science is often considered equivalent to the achievements of **Newton** in the field of gravity. **Faraday** had already discovered that there is some kind of deep connection between electricity and magnetism, showing that a changing electric current sets up a magnetic field, which can in turn induce a current in a nearby wire. Maxwell established the relationships between electricity and magnetism with firm mathematical foundations for the first time.

His four famous equations represent simple and elegant ways to state the relationships between electric and magnetic fields, in terms of variables such as magnetic and electric field strength as well as constants, including the 'electric permittivity of free space' and the 'magnetic permeability of free space'. Another constant that cropped up in the equations was particularly interesting. Maxwell discovered that electric oscillations produce a magnetic field that expands outward at a constant speed of about 300,000 km (186,300 miles) per second, close to the speed of light. He quickly made the link that light itself is a form of electromagnetic radiation.

His equations first appeared in a fully developed form in the textbook *A Treatise on Electricity and Magnetism* (1873). Although it was already known that light can have wavelengths beyond those that the human eye can detect (such as infrared and ultraviolet light), Maxwell's theory proved for the first time that there should be wavelengths well beyond these limits. Sure enough, short-wavelength electromagnetic radiation such as X-rays and gamma rays, and long-wavelength radio waves, were later discovered.

At the time, Maxwell believed that light required a medium (often called the 'ether') in which to propagate, just as sound waves require a medium such as air. However, this idea gradually became discredited, especially following a famous experiment to test the idea by **Michelson** and his colleague Edward Morley. The constancy of the speed of light in a vacuum – regardless of the speed of the source emitting it – became a central tenet of **Einstein**'s special relativity theory.

Maxwell's quantitative connection between light and electromagnetism is often considered to be the greatest accomplishment of 19th-century mathematical physics. In *The Feynman Lectures on Physics* (1964), **Feynman** wrote: 'From a long view of the history of mankind – seen from, say, ten thousand years from now – there can be little doubt that the most significant event of the 19th century will be judged as Maxwell's discovery of the laws of electrodynamics. The American Civil War will pale into provincial insignificance in comparison with this important scientific event of the same decade.' Maxwell died young from cancer at the age of 48.

Scottish–American scientist and inventor who designed the first practical telephone

'I then shouted into [the mouth piece] the following
sentence: "Mr Watson – come here – I want to see you."
To my delight he came and declared that he had heard
and understood what I said.'

Bell describes his first successful phone conversation with his assistant
Thomas Watson on 10 March 1876 (from Bell's experimental notes)

Bell was born in Edinburgh to a mother who was becoming increasingly deaf, and his father and grandfather were both speech therapists, factors that enormously influenced his future work. Although as a child he was interested in science and tinkering with mechanical devices, he didn't excel at school and left aged 15, when he went to live and study with his grandfather in London.

After teaching elocution briefly in Elgin, Scotland, he attended the University of Edinburgh and University of London. He emigrated with his family to Canada in 1870 following the death of his two brothers from tuberculosis. In Canada, Bell set up a workshop to study electricity and sound, and in 1871 he moved to Boston, Massachusetts, where he set up a training programme for teachers of the deaf and later a new school for deaf pupils. He later became professor of physiology and elocution at Boston University.

Bell's early experiments included developing a 'harmonic telegraph', a device that allowed multiple messages to be simultaneously transmitted through a wire encoded as different pitches, as well as a 'phonautograph' that could trace vibrations of sound waves and physically draw them onto smoked glass. He became intrigued by the idea that it might be possible to turn sound waves into undulating electric currents that could be converted back into sound waves again using metal reeds tuned to different frequencies.

By 1874, use of the telegraph to transmit text messages using codes such as Morse code was becoming widespread commercially, and there was a huge desire to find ways to send multiple messages over telegraph lines to avoid the expense of constructing new ones. That inspired two wealthy patrons to back Bell's work on sending multiple tones over a telegraph wire. He was awarded the first US patent for the telephone in 1876 and on 10 March in that year, he made his first successful experiment with the telephone, calling his assistant Thomas Watson.

The key to his telephone's success was that vibration of the mouthpiece due to the sound of speech made a needle vibrate in water, which varied electrical resistance in the telephone circuit. This created variable current in the telephone wire, which made an electromagnet vibrate a disc in the receiver, creating sound waves in the air. Other inventors were independently working on telephones, including the American electrical engineer Elisha Gray – some historians credit Gray with the invention, although unlike Bell he failed to secure the patent.

A few months later, Bell successfully operated a telephone over several miles. In 1877, he founded the Bell Telephone Company to commercialize the phone; Bell owned a third of the shares and quickly became a wealthy man. He became a naturalized citizen of the US in 1882.

Bell was also controversially linked to the eugenics movement. In 1921, he was honorary president of a eugenics congress hosted by the American Museum of Natural History in New York and he advocated sterilization of people who are a 'defective variety of the human race', including deaf people. Although this view sounds extreme today, it was common at the time.

THOMAS ALVA EDISON
11 February 1847 – 18 October 1931

Prolific American inventor who introduced a host of devices including long-lasting light bulbs

'None of my inventions came by accident. I see a worthwhile need to be met and I make trial after trial until it comes. What it boils down to is 1 per cent inspiration and 99 per cent perspiration.'

Stated by Edison in a press conference in 1929 (quoted in *Uncommon Friends: Life with Thomas Edison, Henry Ford, Harvey Firestone, Alexis Carrel & Charles Lindbergh* by James Newton, 1989)

From the age of seven, Edison was brought up in Port Huron, Michigan. An inquisitive child, he began experimenting at an early age but received almost no formal education. From 1862, he worked as a telegrapher and became familiar with electrical equipment by experimenting with telegraph technology. In 1868, he became an independent inventor in Boston.

In 1869, Edison moved to New York, where he was contracted by telegraph companies to develop the technology. The financial income allowed him to set up manufacturing shops in New Jersey and he soon developed a reputation as a prolific inventor. He designed fire alarms, for instance, and made improvements to the 'stock ticker', an ancestor of the modern computer printer.

In 1876, Edison created the first industrial research centre with a machine shop and laboratories at Menlo Park (now Edison), New Jersey, on the railway line between New York City and Philadelphia. He eventually employed about 80 scientists. Urged by Western Union to develop a telephone that could compete with **Alexander Graham Bell**'s, Edison invented a microphone in which the vibrations of a user's voice changed the resistance of a carbon component, a new principle used in telephones for the following century.

Around this time, Edison also unveiled the phonograph, which recorded sound on tinfoil around a grooved cylinder. His next great innovation was to improve the electric light bulb. After thousands of experiments, he finally created a high-vacuum bulb with a cotton filament that could shine for 40 hours. In 1887, Edison built an industrial research laboratory in West Orange, New Jersey, and this became the primary research facility for the Edison lighting companies. By 1887, there were 121 Edison power stations in the US delivering DC electricity to customers.

In 1888, Edison came up with the idea of a motion picture device (to do 'for the eye what the phonograph does for the ear') and he filed two preliminary patents, the second one naming the device the 'kinetoscope'. He tasked the inventor William Dickson with turning it into a reality, leading to the first practical celluloid film for movies with the 35 mm standard width often still used today.

During the First World War, Edison led the Naval Consulting Board, which assessed possible inventions for military use including enemy submarine detectors. By the time of his death in 1931, he had received 1,093 US patents and 2,332 patents worldwide, and he remained the world's most prolific inventor on record throughout the 20th century.

LUDWIG EDUARD BOLTZMANN
20 February 1844 – 5 September 1906

Austrian physicist who established classical statistical physics and developed a kinetic theory of heat

'His line of thought may be called my first love in science. No other has ever thus enraptured me or will ever do so again.'

Erwin Schrödinger enthuses about Boltzmann's statistical methods in the essay collection *Science Theory and Man* (1957 reprint)

Boltzmann studied at the university in his home town, Vienna, where he earned his doctorate in 1866 for a thesis on the kinetic theory of gases. He later studied under **Bunsen** in Heidelberg, and under **Kirchhoff** and **Helmholtz** in Berlin. In 1869, Boltzmann became professor of theoretical physics at Graz, and he later held professorships in Vienna and Leipzig.

During the 1860s, atomic theory was still highly controversial; it was still common to believe that matter could be divided indefinitely. However, **Maxwell** and others were rapidly developing kinetic theories of gases, which describe a gas as a large number of tiny particles bouncing off each other in constant, random motion. Boltzmann extended this theory, showing mathematically how kinetic energy is divided between the gas particles at any one time depending on factors like the bulk temperature of the gas.

In other work, Boltzmann developed a statistical interpretation of the second law of thermodynamics. He also provided a mathematical explanation for why the thermal radiation emitted by a hot body (all normal matter emits radiation because it is warm) is proportional to the fourth power of its temperature (T^4).

Boltzmann suffered extreme mood swings, possibly due to bipolar disorder, and struggled to cope with criticism from scientists who maintained an anti-atomist standpoint. In 1906, he hanged himself while on holiday with his wife and daughter near Trieste in Italy.

LUTHER BURBANK
7 March 1849 – 11 April 1926

American horticulturist who pioneered agricultural science, developing more than 800 new plants

'What a joy life is when you have made a close working partnership with Nature, helping her to produce for the benefit of mankind new forms, colours and perfumes in flowers which were never known before; fruits in form, size and flavour never before seen on this globe; and grains of enormously increased productiveness ... new food for all the world's untold millions for all time to come'

From a speech Burbank gave to the First Congregational Church of San Francisco in 1925 (*Mendel in the Kitchen* by Nina Fedoroff and Nancy Marie Brown, 2004)

The 13th of 15 children, Luther Burbank was born in Lancaster, Massachusetts. He grew up on a farm and never had a university education. But even as a child, he had a knack for growing and cross-breeding plants and from reading **Darwin**'s works, he learned the importance of variation.

His botanical work began in earnest when he bought a plot of land in Massachusetts in 1870. There he developed the Burbank potato, a blight-resistant version that was later introduced to Ireland to prevent famines like the one that killed around a million people during 1845 to 1852. A variety of Burbank potato remains the most widely cultivated potato in the US today.

After selling the rights to his potato for $150, Burbank moved to California in 1875 and set up a greenhouse and nursery. From 1904, he received several grants for cross-breeding research from the Carnegie Institution of Washington, founded in 1902 to enable scientists with extraordinary vision to work at the cutting edge of their fields. He created hundreds of new varieties of fruits, as well as vegetables, flowers, grains and grasses.

(HEINRICH HERMANN) ROBERT KOCH
11 DECEMBER 1843 – 27 MAY 1910

German physician who founded modern microbiology and isolated the bacterium that causes tuberculosis

'Twenty years ago, tuberculosis, even in its most dangerous form, consumption, was still not considered infectious ... But it was only with the discovery of the tubercle bacillus that the aetiology of tuberculosis was placed on a firm footing, and the conviction gained that this is a parasitic disease, i.e. an infectious, but also avoidable one ... The struggle against tuberculosis is not dictated from above, and has not always developed in harmony with the rules of science, but it has originated in the people itself, which has finally correctly recognized its mortal enemy ... If the work goes on in this powerful way, then the victory must be won.'

From Koch's Nobel lecture (12 December 1905)

Koch was born at Clausthal in the Harz Mountains and studied medicine at the University of Göttingen from 1872, earning his MD degree in 1866. He studied under **Wöhler** and the German physician Jacob Henle, who was an early proponent of the idea that living parasitic organisms cause infectious diseases. He spent several years in general practice and in 1870 he volunteered for service in the Franco–Prussian war before becoming District Medical Officer for Wöllstein from 1872 to 1880. In 1885, he was appointed as professor of hygiene at the University of Berlin.

Koch's international renown grew after his work at Wöllstein, where he studied anthrax. The anthrax bacillus had already been discovered, but Koch was first to demonstrate conclusive proof that it is the cause of the disease. Using slivers of wood, he inoculated mice with anthrax bacilli from the spleens of cows that had died of anthrax and found that these mice all died of anthrax, while mice inoculated at the same time with blood from the spleens of healthy animals did not. This confirmed that the disease can be transmitted by a diseased animal's blood.

At the time, there was an outstanding theoretical question – is it possible for bacteria that have never been in contact with animals to cause disease? To test this, Koch obtained pure cultures of anthrax bacilli by growing them on the aqueous humour of ox's eyes. He monitored the bacilli multiplying and noticed that when conditions are unfavourable to them, they create rounded spores that protect them from adverse conditions, such as a lack of oxygen. When a beneficial environment was restored, the spores gave rise to bacilli again. Koch grew the bacilli for several generations in these pure cultures and showed that even when they had no contact with animals whatsoever, they could still cause anthrax.

Koch's work in observing and photographing disease-causing bacteria helped establish a practical and scientific basis for the control of these infections. He developed new methods of staining bacteria using aniline dyes developed since **Perkins'** time, which made them easier to identify. He developed solid gels for culturing bacteria that effectively confined different generations of a single family of bacteria in the same place, and improved experimental observations of them. He also laid down conditions, known as Koch's postulates, that must be satisfied before scientists jump to the conclusion that a particular bacterium causes disease.

Koch discovered the bacterium that causes tuberculosis and published a classic work on it in 1882. He also searched for a cure for this dreaded disease, but did not find one. In 1883, he travelled to Asia to study bubonic plague and cholera, and he isolated the bacterium that causes cholera (*Vibrio cholerae*). After studying its epidemiology, he promoted public health measures to prevent the spread of the bug. He also studied sleeping sickness in Africa, proved that the disease is transmitted by the tsetse fly and proposed new disease control methods that targeted the insect vector rather than the pathogenic bacteria that they carry.

In later life, Koch came to the conclusion that the bacilli that caused human and bovine tuberculosis are not identical. His work on typhus led to the new notion that it is easily transmitted between people, rather than arising solely from contamination of drinking water (person-to-person transmission is now known to be via body lice). For his work on tuberculosis, Koch won the 1905 Nobel prize for physiology or medicine. He mentored and inspired many excellent bacteriologists of the next generation including **Ehrlich**, and with **Pasteur**, he is considered to be a founder of modern medical bacteriology.

WALTHER FLEMMING
21 April 1843 – 4 August 1905

German biologist who discovered chromosomes and stages of cell division

'We will designate as chromatin that substance, in the nucleus, which upon treatment with dyes known as nuclear stains does absorb the dye ... In nuclear division it accumulates exclusively in the thread figures.'

Flemming describes his observations of chromosomes ('Contributions to the Knowledge of the Cell and its Vital Processes, Part II')

Flemming was born in Schwerin, Mecklenburg, and trained in medicine at the University of Rostock, graduating in 1868. He then became a military physician and served in the Franco-Prussian War of 1870. He later became professor at the University of Prague and at the University of Kiel from 1876 until his death.

By the 1870s, scientists were starting to use synthetic dyes developed by **Perkin** and others to stain transparent cells and see their structures more clearly under the microscope. Using aniline dyes, Flemming found that a network of material in the nuclei of animal cells, which he called chromatin (Greek for 'colour'), preferentially absorbed the dye. He noticed that as cell division began, the chromatin converged into thread-like structures, now called chromosomes.

This striking feature inspired him to coin the term 'mitosis' (from the Greek for 'thread') for the cell division process. He also noted that chromosome numbers doubled as a cell divided, then split apart to create two daughter cells with the same original number of chromosomes. However, he did not identify this process as a mechanism for genetic inheritance. The Dutch geneticist Hugo Marie de Vries made this link two decades later when he rediscovered **Mendel**'s work on heredity.

EMIL KRAEPELIN
15 February 1856 – 7 October 1926

German psychiatrist often considered the founder of modern scientific psychiatry

'A number of patients have been referred to me whose deep sadness, paucity of speech and anxious tension might have suggested a circular depression, yet it came out subsequently that we were dealing with dysphorias caused by serious mistakes [the patients had made] and by looming legal measures ... The milder depressions of manic-depressive illness, as much as we can tell, fully resemble the motivated dysphorias of healthy psychic life – with the essential difference that they occur without motivation ... one will not be able to interpret the symptoms correctly without knowledge of the patient's history.'

Written by Kraepelin in 1913
(*Psychotic Depression* by Conrad Swartz and Edward Shorter, 2007)

Kraepelin studied medicine in Leipzig and Wurzburg and earned his MD in 1878. From 1882, he held a research post at the University of Leipzig, where he worked on psychopharmacology with **Wundt**. In 1884 he became a senior physician in Leubus and in 1885 he was appointed director of the Treatment and Nursing Institute in Dresden. After four years as professor at the University of Dorpat (now Tartu in Estonia), he held a post at the University of Heidelberg until 1904.

Kraepelin wrote an impressive work (*Compendium of Psychiatry*, 1883) arguing that psychiatry is a science that, like any other, should be rigorously tested by experiment. He also developed a logical classification system for mental disorders. He divided cases of psychosis into the classes of manic depression and 'dementia praecox' (effectively schizophrenia).

Crucially, he also showed that both diseases have some kind of inherited and/or environmental link, because they are more common in some families than in the general population. Kraepelin was convinced that biological disorders of the brain cause major psychiatric conditions, and his laboratory uncovered the pathological basis of Alzheimer's disease.

SVANTE AUGUST ARRHENIUS
19 FEBRUARY 1859 – 2 OCTOBER 1927

Swedish scientist who founded the science of physical chemistry

'Chemistry works with an enormous number of substances, but cares only for some of their properties; it is an extensive science. Physics on the other hand works with rather few substances, such as mercury, water, alcohol, glass, air, but analyses the experimental results very thoroughly; it is an intensive science. Physical chemistry is the child of these two sciences.'

From Arrhenius's *Theories of Solutions* (1912)

Born near Uppsala in Sweden, Arrhenius studied sciences at the University of Uppsala and from 1881 at the Physical Institute of the Swedish Academy of Sciences in Stockholm. In 1895, he was appointed professor at the University of Stockholm and ten years later he became head of a new Nobel Institute for Physical Chemistry in Stockholm.

During his PhD work, Arrhenius studied electrolytes (such as sodium chloride solution) and showed that they contain electrically positive and negative ions that can carry electric current. **Faraday** had already suggested that ions arise in the process of electrolysis, due to the passage of electrical current. Arrhenius correctly recognized that even without a current, salt solutions contained charged ions. At the time, it was radical to suggest that mere water could break up a stable substance like common salt.

His PhD thesis barely scraped a pass in 1884 from unimpressed examiners. But for the same theory he won the Nobel prize for chemistry in 1903. By this time, **J.J. Thomson** had discovered the electron, proving that atoms are made of electrically charged particles, so the idea of ions followed naturally. Arrhenius also formulated the concept of activation energy – the energy, such as heat, needed to make molecules react with each other. His simple Arrhenius equation defines how activation energy and temperature determine chemical reaction rates.

ALBERT ABRAHAM MICHELSON
19 December 1852 – 9 May 1931

American physicist who developed precision optical instruments and carried out the famous Michelson–Morley experiment

'Professor Michelson's brilliant adaptation of the laws of light interference has ... perfected a group of measuring instruments, the so-called interferometers ... an increase in accuracy in measurement of from twenty to a hundred times what can be achieved with the best microscopes has been brought well within our grasp'

From the presentation speech written for Michelson's Nobel prize (by Bernhard Hasselberg, Royal Swedish Academy of Sciences, 10 December 1907)

Michelson grew up in San Fransisco after his family emigrated to the US when he was two. He studied at the US Naval Academy, and later held professorships at the Case School of Applied Science in Ohio (now Case Western Reserve University) and the University of Chicago before joining Mount Wilson Observatory in California in 1929.

Michelson excelled at making precision optical instruments and made early accurate measurements of the speed of light. But he is best known for a famous experiment he performed with Edward Morley in Ohio in 1887. At the time, there was heated debate about whether or not light waves required a medium – the hypothetical 'ether' – in which to propagate, just as sound waves require a physical medium. If so, the speed of light should depend on propagation direction, because of the Earth's orbital motion relative to the ether. The experiment showed that the ether doesn't exist – light travels at a constant speed regardless of the Earth's motion.

For his work on optics, Michelson won the 1907 Nobel prize for physics. In astronomy, he also made the first accurate determination of the size of a star (Betelgeuse) beyond the Sun.

(JULES) HENRI POINCARÉ
29 April 1854 – 17 July 1912

*French mathematician who laid the foundations for
Einstein's special theory of relativity*

'The scientist does not study nature because it is useful to
do so. He studies it because he takes pleasure in it, and he
takes pleasure in it because it is beautiful ...
I am not speaking, of course, of the beauty which strikes
the senses, of the beauty of qualities and appearances ...
What I mean is that more intimate beauty which comes
from the harmonious order of its parts, and which a
pure intelligence can grasp.'

From Poincaré's *Science and Method* (1908)

Born in Nancy in northeast France, Poincaré earned a degree
in letters and science at the Lycée in Nancy in 1871. From 1873, he studied
maths at the École Polytechnique in Paris. He went on to study maths and
mining engineering at the École des Mines and earned a doctorate from
the University of Paris in 1879. He later held several engineering posts and
taught sciences at the Sorbonne from 1881 till the end of his career.

Poincaré made wide-ranging contributions to science and maths.
He formulated the Poincaré conjecture, a mathematical assertion that
remained unproven until 2003 (see **Perelman**). He also developed
the principle of relativity, that all observers must experience the same
physical laws regardless of their relative motion. This is a cornerstone of
Einstein's special theory of relativity, for which Poincaré laid much of the
mathematical groundwork.

In addition, he made key advances in solving the 'three-body problem',
the quest to find a general solution to the motion of three orbiting bodies,
which had eluded mathematicians since **Newton**'s time. Among other
topics, he studied Saturn's rings and the stability of the Universe, and he
wrote extensively on philosophy.

JAMES DEWAR
20 September 1842 – 27 March 1923

Scottish chemist and physicist who pioneered gas liquefaction and invented the vacuum flask

'Minds are like parachutes, they only function when they are open'

Quote often attributed to Dewar

Dewar was born in Kincardine, Firth of Forth. He studied at the University of Edinburgh and later under **Kekulé** at the University of Ghent. In 1875, he became Jacksonian Professor of Natural Philosophy at the University of Cambridge, a post he held until his death. From 1877, however, he was also Fullerian Professor of Chemistry at the Royal Institution in London, where he carried out most of his experiments.

In 1867, Dewar described several possible chemical formulae for benzene and in 1889, he co-invented the smokeless gunpowder cordite. But he is probably best known for his experiments on cooling and liquefying gases, including oxygen and air. In the late 1890s, he also succeeded in liquefying hydrogen by cooling the gas at high pressure to about −200°C.

By 1891, Dewar had designed and built machinery to produce industrial quantities of liquid oxygen. Soon after that, he invented the 'Dewar flask' with an outer vacuum chamber for storing the liquid gases. He did not patent his invention, which was marketed as the Thermos flask from the early 1900s. Dewar studied many other topics including the temperature of the Sun and the properties of fluorine, as well as the behaviour of soap bubbles, and he was knighted in 1904.

(HERMANN) EMIL FISCHER
9 October 1852 – 15 July 1919

*German organic chemist who synthesized sugars
and deduced their structures*

'Progressively, the veil behind which Nature has so carefully
concealed her secrets is being lifted where the carbohydrates are
concerned. Nevertheless, the chemical enigma of Life will not be
solved until organic chemistry has mastered another, even more
difficult subject, the proteins, in the same way as it has mastered
the carbohydrates ... On this hard ground the fruit ripens far more
slowly and the total amount of work that has to be done here is
so enormous that in contrast the elucidation of the carbohydrates
seems child's play.'

From Fischer's Nobel lecture (12 December 1902)

Fischer grew up being encouraged to join
his family's business in Euskirchen, Prussia (now in
Germany). However, he wasn't cut out for the job and
instead pursued his stronger interest in science. From
1871, he studied chemistry at the University of Bonn,
where he attended lectures by **Kekulé**. The following
year he moved to the newly established University of
Strassburg, where he earned a PhD in 1874.

Fischer was then appointed as an instructor at the same university.
He later held posts at the University of Munich, where he was appointed
as associate professor in 1879, and the University of Erlangen, where
he became professor in 1881. He was later professor at the University of
Würzburg then Berlin University, where he remained until his death.

In his early work, Fischer discovered the first hydrazine base,
phenylhydrazine ($C_6H_5NHNH_2$). At Erlangen, he studied chemicals in
tea, coffee and cocoa, including caffeine and theobromine, and eventually
he developed methods to synthesize them. But he is best known for his
research on purines and sugars. In this work, carried out between 1882 and
1906, he showed that adenine, xanthine, caffeine, uric acid and guanine all
belong to one chemical family and can be derived from one another.

Fischer named the parent substance for these products purine ($C_5H_4N_4$) in 1884 and he went on to synthesize it in 1898. His laboratory created many artificial derivatives of purine between 1882 and 1896. This laid important foundations for later developments in genetics – two of the four deoxyribonucleotides and two of the four ribonucleotides, the respective building blocks of DNA and its molecular cousin RNA, are purines.

In 1884, Fischer began his extensive work on sugars, which created a coherent picture of their structures for the first time. In 1890, he showed that there are two series of sugars that are stereoisomers – they have exactly the same molecular formula but form mirror images of each other. He called one the D-series and one the L-series.

One of Fischer's greatest successes was to synthesize glucose, fructose and mannose in 1890, starting from glycerol. Between 1899 and 1908, he also clarified the structures of proteins. It was already known that the building blocks of proteins are amino acids, but Fischer showed exactly how these are combined within proteins. He developed effective methods for separating and identifying individual amino acids, and discovered a new type, the cyclic amino acids proline and oxyproline. He showed that they can be connected in chains by peptide bonds and went on to produce dipeptides, tripeptides and polypeptides.

Fischer also studied carbohydrates, fats and the enzymes in lichens, as well as substances used in tanning. For his work on sugar and purine synthesis, he won the 1902 Nobel prize for chemistry. He was passionately committed to basic research and campaigned for the creation of new research institutes, gathering support for the establishment of the Kaiser Wilhelm institutes for chemistry and coal research. In Berlin, he also organized a radiochemistry laboratory where **Hahn** and **Meitner** would later discover nuclear fission.

During the First World War, Fischer worked with the German government and industry to synthesize chemicals for explosives that had become unavailable due to the British blockade. His life was overshadowed by personal tragedy, however. His wife died after just seven years of marriage and one of his three sons was killed during the First World War. A second son committed suicide. In 1919, having developed cancer, Fischer himself committed suicide at the age of 66.

NIKOLA TESLA
10 July 1856 – 7 January 1943

*Serbian–American electrical engineer whose work
ushered in the age of commercial electricity*

'The first question to answer then is whether pure
resonance effects are producible. Theory and experiment
show that such is impossible in nature for, as the
oscillations become more vigorous, the losses in vibrating
bodies and environing media rapidly increase, and
necessarily check the vibrations, which would otherwise
go on increasing forever. It is a fortunate circumstance
that pure resonance is not producible, for, if it were,
there is no telling what dangers might lie in wait for the
innocent experimenter.'

Tesla considers the power of 'resonance', a phenomenon that fascinated him
(*The Man who Invented the Twentieth Century* by Robert Lomas, 1999)

Tesla was born into a Serbian family in
the village of Smiljan in the Austrian Empire (now in
Croatia). He studied engineering at Graz Polytechnic
and the University of Prague. After that he worked
for a company that installed telephone systems
in Budapest, where he designed components
such as generators, motors and transformers for
long-distance distribution of power as alternating
current at high voltages, a much more efficient way of distributing power
than via direct current.

From 1882, he worked for the Continental Edison Company in Paris,
using innovations designed by **Edison** to improve electrical equipment.
In 1884, he emigrated to the US, where he contacted Edison in New
Jersey. They had a brief collaboration, but Tesla severed contact with
Edison following a dispute over payment. Tesla then established his own
laboratory in New York City in 1887.

Tesla's alternating-current system for efficient power distribution
was adopted for major energy projects from the 1890s. He was by then
churning out a stream of ingenious electrical inventions for power

distribution, and his patents and theoretical work also formed the basis of wireless communication and the radio. He also invented fluorescent lights, a new type of steam turbine and, bizarrely, a vibrating platform for use as a mechanical laxative.

In 1891, Tesla became a naturalized citizen of the US. Around this time, he invented the transformer circuit now known as the Tesla coil, which relies on electrical resonance. Tesla was mesmerised by resonance, the notion that small forces, if carefully timed, can add up to create huge energies, just as a child on a playground swing goes higher and higher when given a small push at the right time. While testing a vibration machine in his lab one day, he hit the natural frequency of vibration of the four-storey building and convinced his terrified neighbours that they were about to die in a devastating earthquake.

The Tesla coil – actually two coils, one nested inside the other – exploits electrical resonance within the coils. When an alternating current builds in the primary coil, it generates a magnetic field that induces current in the secondary coil. If both coils have the same resonant frequency, the voltage on the secondary coil can climb to enormous values and the system effectively converts high current and low voltage in the primary into a low current and high voltage in the secondary. The voltage can soar so high that it breaks down the air around it in a crackling lightning-like discharge.

In 1901, Tesla began constructing Wardenclyffe Tower, a giant Tesla coil 57 m (187 ft) high on the coast of Long Island. He had described the project to financiers as a high-power radio transmitter for communications, but secretly, he hoped to use it for wireless transmission of electrical power using the Earth's ionosphere as an electrical circuit. When the chief investor realized this he pulled the plug on the project. The tower was eventually destroyed during the First World War to prevent German submarines using it as a landmark for navigation.

Tesla was a lonely, reclusive and eccentric character who had a poor head for business and latterly became prone to making long rambling speeches. He spent the last decade of his life living in a New York hotel, where he died. By that time, he had been granted more than 700 patents. Around 160,000 of his original documents are today housed at the Nikola Tesla Museum in central Belgrade and the standard international unit of magnetic flux density, the tesla, was named in his honour.

WILLIAM RAMSAY
2 October 1852 – 23 July 1916

Scottish chemist who discovered the noble gases

'Helium, like argon, is a gas, sparingly soluble in water,
withstanding the action of oxygen in presence of caustic soda,
under the influence of the electric discharge, as well as of
red-hot magnesium ... These properties in common made
it evident that helium and argon belong to the same natural
family; and it was also obvious that there must exist at least
three other elements of the same class; this is evident
on inspection of the periodic table.'

From Ramsay's Nobel lecture (12 December 1904)

Ramsay's paternal grandfather (also called William Ramsay) was a chemical manufacturer who came from a long line of dyers and his maternal grandfather wrote chemistry textbooks for medical students, and Ramsay inherited a taste for chemistry from both sides of his family. However, he opted to first study classics at the University of Glasgow. He later studied chemistry at Tübingen in Germany, where he earned his doctorate in 1872.

Following his return to Scotland, Ramsay became an assistant in chemistry at Anderson's College in Glasgow. In 1880, he became professor of chemistry at University College, Bristol, then in 1887 he was appointed as professor of inorganic chemistry at University College London, a post that he held until his retirement in 1913.

Ramsay's earliest work was on organic chemistry, including a study of the decomposition products of the alkaloid quinine, used at the time as a treatment for malaria. But it was in inorganic chemistry that he made most of his celebrated discoveries. From 1892, he investigated a puzzle highlighted by the English physicist John William Strutt (Lord Rayleigh), who had noted that nitrogen derived from air always had a slightly higher atomic weight than nitrogen released by controlled chemical reactions in the lab.

Ramsay's hunch was that the Earth's atmosphere contains traces of a previously unidentified gas that contaminated the atmosphere-derived nitrogen and had a higher atomic weight. To test this, he took a sample of air, removed the oxygen through combustion reactions and then removed the nitrogen by reacting it with hot magnesium. This left behind traces of an inert gas that was later named argon.

In 1895, he went on to discover helium, with similar properties to argon, after boiling a mineral called cleveite. Helium had already been discovered through spectroscopic studies of the Sun as well as volcanic lava, but Ramsay was the first person to isolate the gas. He also correctly placed argon and helium in the periodic table of the elements, filling gaps that **Mendeleyev** had highlighted. Realizing that there were further gaps in the inert gas family, Ramsay carried out spectroscopic studies of air from which the oxygen and nitrogen had been removed. This led to his discovery of neon, krypton and xenon.

He also showed that the radioactive gas radon, discovered in 1900, is a member of the inert gas family. Working with the English radiochemist Frederick Soddy, he confirmed in 1903 that the radioactive decay of radium produces alpha particles composed of positively charged helium nuclei. This work would later prove a key milestone in clarifying how nuclear reactions occur.

For discovering the inert gases and determining their place in the periodic table, Ramsay won the 1904 Nobel prize for chemistry. The discoveries had required immense precision due to the tiny amounts of inert gases in the atmosphere, something that Ramsay remarked on in his Nobel lecture: 'The amounts of neon and helium in air have since been measured; the former is contained in air in the proportion of one volume in 81,000; the latter, one volume in 245,000; the amounts of krypton and xenon are very much smaller – not more than one part of krypton by volume can be separated from 20,000,000 of air; and the amount of xenon in air by volume is not more than one part in 170,000,000.'

Ramsay was also knighted in 1904, and Rayleigh won the Nobel physics prize in the same year for his work on the densities of gases, which led to the discovery of argon.

PERCIVAL LAWRENCE LOWELL
13 March 1855 – 12 November 1916

*American astronomer who predicted the existence
of a planet beyond Neptune*

'With regard to the observatory's plan of work, its main
object is the study of our own solar system. This may
be put popularly as an investigation into the condition
of life in other worlds, including last but not least, their
habitability by beings like [or] unlike man.'

From Lowell's address to the Boston Scientific Society on setting
up the first observatory sited in a remote, high place for
improved observing conditions (1894)

Born into a wealthy and influential family
in Boston, Massachusetts, Lowell graduated from
Harvard University in 1876 with distinction in
mathematics. After that he travelled in the Far East.
On his return to the US, he founded the Lowell
Observatory just west of Flagstaff, Arizona, which
opened in 1894.

Lowell spent many years at the observatory
photographing and drawing what he believed to be linear Martian canals,
which he saw as evidence of intelligent life on Mars. He also noticed
seasonal changes on the Martian surface, which he thought represented
agricultural progression. It is likely that many of the features he described
were optical illusions seen through his relatively primitive telescope.

Lowell is best known for his prediction that a planet exists beyond
Neptune. He believed the orbits of Uranus and Neptune were disturbed by
the gravity of a more distant body, and he searched for this 'Planet X' from
1906. **Tombaugh** eventually discovered a planet beyond Neptune (Pluto,
now classified as a dwarf planet) in 1930, more than a decade after Lowell
died. Lowell's prediction fuelled the search for the new planet, but modern
data on the planets remove the need for Planet X as Lowell defined it.

IVAN PETROVICH PAVLOV
14 September 1849 – 27 February 1936

Russian physiologist famous for his 'Pavlov's dog' experiments that revealed conditioned reflexes

'Essentially only one thing in life interests us: our psychical constitution, the mechanism of which was and is wrapped in darkness. All human resources, art, religion, literature, philosophy and historical sciences, all of them join in bringing light in this darkness. But man has still another powerful resource: natural science with its strictly objective methods. This science, as we all know, is making huge progress every day. The facts and considerations which I have placed before you at the end of my lecture are one out of numerous attempts to employ a consistent, purely scientific method of thinking in the study of the mechanism of the highest manifestations of life in the dog, the representative of the animal kingdom that is man's best friend.'

From Pavlov's Nobel lecture (12 December 1904)

Pavlov was born in Ryazan, southeast of Moscow. After theological studies, he studied natural sciences at the University of St Petersburg, graduating in 1875, and medicine, receiving his doctoral degree in 1879. In 1890, he became head of the physiology department at the newly founded Institute of Experimental Medicine in St Petersburg.

Pavlov is famous for the discovery of the 'conditioned reflex' – that an environmental stimulus that previously had no relation to a reflex could through repeated experience trigger that reflex. In dogs, he showed that if he rung a bell whenever he gave food to a dog, the dog would eventually salivate when the bell rang, even if no food was on offer. Experiments by Pavlov and his students showed that conditioned reflexes originate in the brain's cerebral cortex. For this work, Pavlov won the 1904 Nobel prize for physiology or medicine.

WILHELM CONRAD RÖNTGEN
27 March 1845 – 10 February 1923

German physicist who discovered X-rays

'The effect was one which could only be produced, in ordinary parlance, by the passage of light. No light could come from the tube, because the shield which covered it was impervious to any light known, even that of the electric arc ... I assumed that the effect must have come from the tube, since its character indicated that it could come from nowhere else. I tested it. In a few minutes, there was no doubt about it. Rays were coming from the tube which had a luminescent effect upon the paper. I tried it successfully at greater and greater distances, even at two metres. It seemed at first a new kind of invisible light. It was clearly something new, something unrecorded.'

From an interview with Röntgen by H.J.W. Dam
(published in *McClure's Magazine*, April 1896)

Röntgen was the only child of a cloth manufacturer and merchant in Lennep, born in the Lower Rhine Province. His family moved to the Netherlands when he was a young child and he studied physics at the University of Utrecht from 1865. He later studied mechanical engineering at the Polytechnic at Zurich and in 1869 earned a PhD from Zurich University. He later held professorships in Württemberg, Strassburg, Giessen and Munich.

At one point he planned to emigrate to the US and accepted a post at Columbia University in New York City. However, following the outbreak of the First World War he decided against the move and remained in Munich until his retirement.

In early work, Röntgen studied the specific heats of gases and the thermal conductivity of crystals. But he is best known for discovering X-rays. In 1895, he studied the effects of passing an electric current through a gas of extremely low pressure in a cathode-ray tube, and noticed that even if the discharge tube was enclosed in a sealed, thick black carton to exclude all light, the chemical barium platinocyanide placed nearby became fluorescent. In rigorous experiments to investigate this further, he showed that objects with increasing thickness blocked this effect to increasing degrees when recorded on a photographic plate.

Röntgen exposed his wife's hand to the rays above a photographic plate and observed an image of her bones and a ring on her finger when he developed the plate. He didn't understand the nature of the rays that were causing this effect, and named them X-rays. He recognized that the discovery could be ground-breaking and continued experiments on X-rays in secret, fearing that his professional reputation could be destroyed if his claims turned out to be false. But by late 1895, he was convinced that he was right. At that point, he submitted a paper on the subject for publication, demonstrating not only the discovery of X-rays but many of their properties, such as the ability to ionize gases and their lack of deflection in electric or magnetic fields, which made it unlikely that they were composed of charged particles.

In total, Röntgen published three papers on X-rays between 1895 and 1897. Today, he is considered the founder of diagnostic radiology, which would dramatically improve techniques for medical diagnosis. Later, **von Laue** and his students showed that X-rays are of the same electromagnetic nature as visible light, but have a very high frequency of oscillation that makes them invisible to the human eye.

Röntgen won the Nobel prize for physics in 1901, the first year that the prize was awarded, for 'the extraordinary services he has rendered by the discovery of the remarkable rays'. His research also included work on elasticity, capillary movement of fluids, and piezoelectricity, the effect in which charge accumulates in certain solid materials such as quartz crystals in response to mechanical stress applied across them. A unit of measurement for exposure to ionizing radiation, the röntgen, was named in his honour, as was roentgenium, a radioactive chemical element with the atomic number 111.

Despite his fame and the numerous honours bestowed on him during his lifetime, Röntgen remained a very modest and reticent character. He died in 1923 from cancer of the intestine. While exposure to ionizing radiation is thought to have contributed to the deaths of early pioneers of radiation experiments, including **Marie Curie**, this was probably not the case for Röntgen. His experiments with X-rays were relatively short lived and he used protective lead shields in many of his tests.

(ANTOINE) HENRI BECQUEREL
15 December 1852 – 25 August 1908

French physicist who co-discovered radioactivity

'I soon recognized that the emission was independent of any
familiar source of excitation, such as light, electricity or heat.
We were thus faced with a spontaneous phenomenon
of a new order.'

Becquerel describes his discovery of radioactivity in his Nobel lecture (11 December 1903)

Paris-born Becquerel studied science at the prestigious École Polytechnique in the city then later engineering at the École des Ponts et Chaussées (School of Bridges and Roads). In 1892, he became professor of physics at Paris's Museum of Natural History and in 1894, chief engineer for the French department of bridges and highways.

The discovery that made Becquerel famous came in 1896. By then, he had heard about the radiation (X-rays) discovered by **Röntgen**, and he wondered if there was any link between X-rays and naturally occurring phosphorescence (from glow-in-the-dark materials). He had inherited from his father a supply of uranium salts, which phosphoresce after exposure to light. If the salts emitted penetrating X-rays, Becquerel reasoned, Röntgen's results suggested this would fog a photographic plate even if the plate was covered with opaque paper.

Sure enough, that did happen. But Becquerel accidentally discovered the true nature of the 'rays' emitted by uranium when he left a plate in a drawer with uranium crystals that were not phosphorescing because they had not been exposed to light. They nonetheless fogged the plate. Becquerel showed that the rays emitted by uranium, unlike X-rays, could be deflected by electric or magnetic fields – this demonstrated that they were actually charged particles emitted during radioactive decay. For his discovery, Becquerel won half of the 1903 Nobel prize for physics, shared with **Marie** and **Pierre Curie**.

EDMUND BEECHER WILSON
19 October 1856 – 3 March 1939

*American zoologist and cell biologist considered
a founder of modern genetics*

'It has become ever more clearly apparent that the key to all ultimate
biological problems must, in the last analysis, be sought in the cell.
It was the cell theory that first brought the structure of plants and
animals under one point of view by revealing their common plan of
organization ... The cell theory must therefore be placed beside the
evolution theory as one of the foundation stones of modern biology.'

From Wilson's *The Cell in Development and Inheritance* (1896)

Wilson was born in Geneva, Illinois.
He studied biology at Antioch College in Ohio and
at Yale University in Connecticut, and earned a PhD
from Johns Hopkins University in Maryland in 1881.
Following that he spent a year in Europe, where he met
many influential scientists including **Huxley**. After
holding several teaching posts he became professor at
Columbia University in New York from 1891.

With the work of **Virchow** and others, the latter part of the 19th
century had seen the widespread acceptance of the notion that all
living things have cells as their basic unit. In studies of embryos,
Wilson traced how an egg cell divides into two, four, and then eight
and more cells, and established our basic understanding of cell
division. He recognized that these divisions were influenced both
by an inherited tendency and an environment determined by other
surrounding cells.

In insect experiments, Wilson also discovered the role of chromosomes
in sex determination (independently of **Stevens**), showing in 1905 that the
inheritance of the Y chromosome confers male sex. His book *The Cell in
Inheritance and Development* (first edition 1896) is considered a highly
influential classic work on cell biology.

J(OSEPH) J(OHN) THOMSON
18 December 1856 – 30 August 1940

English physicist who discovered the electron

'Not so very long ago the atom was thought to be a terminus beyond which it was impossible from the nature of things to penetrate. The atom was regarded as indivisible, impenetrable, eternal, unaffected by heat, electricity or any other physical agent. The inside of the atom was regarded as a territory which the physicist could never enter. Then there came a time when the sanctuary of the atom was invaded, and it was found that the atom was built up of smaller parts – of electrons carrying a charge of negative, and of protons carrying a charge of positive electricity ... The atom, instead of being just the little hard solid particle of the original view, was a very complex thing, comparable in complexity with the solar system.'

From Thomson's lecture 'Beyond the Electron' (8 March 1928)

Thomson was born in Cheetham Hill, a suburb of Manchester. He enrolled at Owens College (later part of the University of Manchester) in 1870 and from 1876 he studied maths at Cambridge University. From 1883 he was a lecturer at Cambridge and he was appointed as Cavendish Professor of Experimental Physics there from 1884 until 1919. He also served as president of the Royal Society from 1915 to 1920.

Thomson is best known for his experiments in the late 1890s that led to the discovery of the electron. He was working with cathode-ray tubes, in which a high voltage is applied across a vacuum tube, making the negative electrode glow. Some scientists had speculated that the cathode rays were a stream of charged particles. If so, it should be possible to demonstrate that they are deflected by the application of an electric field orientated to make the charged particles curve off their straight path.

Thomson was the first person to demonstrate this by using cathode-ray tubes that were extremely highly evacuated. His experiments also allowed him to calculate the charge-to-mass ratio of the particles. If the charge on the particles was equal to the minimum charge derived from **Faraday**'s laws of electrochemistry, the result implied that the particles were amazingly light – less than a thousandth of the mass of a hydrogen atom. For the first time, Thomson had discovered a subatomic particle that was quickly identified as the negatively charged carrier of electrical current.

He correctly recognized that the electron is a universal component of matter and that the source of the particle stream in the cathode-ray tube was electrons stripped out of the remaining gas in the imperfect vacuum. Thomson envisaged the atom through a 'plum pudding' model, with electrons dotted through a sphere of positive charges like fruit in a Christmas pudding, the total charge on the electrons cancelling the positive charge overall. This would later turn out to be an unrealistic model of the atom when, in 1909, Thomson's student **Rutherford** carried out another famous experiment showing that the positive charge of an atom is concentrated in a dense central nucleus with electrons orbiting around it.

In other experiments, Thomson effectively invented the technique of mass spectrometry. In 1912, he applied electric and magnetic fields to a beam of ionized neon atoms that carried positive charge, and measured the deflection of the beam when it hit a photographic plate. The plate showed not one but two spots where the neon ions arrived, showing that the neon was a mixture of atoms with two different atomic masses.

These turned out to be the isotopes neon-20 and neon-22, the latter having two extra neutrons in its atomic nucleus. Although the English radiochemist Frederick Soddy had already proposed the existence of isotopes to explain patterns of decay in radioactive elements, this was the first evidence that stable elements can also have different isotopes. Thomson's assistant Francis Aston developed this experimental technique further and went on to discover a wide range of isotopes in other elements.

For his discovery of the electron as well as other work, Thomson won the 1906 Nobel prize for physics. He was knighted in 1908 and is buried in Westminster Abbey, close to **Newton**. His son George Paget Thomson also went on to receive a share of the 1937 Nobel prize for physics for the experimental discovery of the diffraction of electrons by crystals, and he was knighted in 1943. Many of Thomson's colleagues also won Nobel prizes, including Rutherford and Aston.

RICHARD DIXON OLDHAM
31 July 1858 – 15 July 1936

Irish geologist who proposed that the Earth has a central core

'Of all regions of the Earth none invites speculation more
than that which lies beneath our feet, and in none is
speculation more dangerous; yet apart from speculation,
it is little that we can say regarding the constitution of the
interior of the Earth ... these waves, penetrating to greater
depths, have entered, and for part of their way traversed,
a central core, composed of matter which transmits them
at a much slower speed'

From Oldham's paper 'The constitution of the interior of the Earth as revealed
by earthquakes' (*Quarterly Journal of the Geological Society*, 1906)

Originally from Dublin, Oldham studied at the Royal School of Mines in London. From 1879 to 1903, he worked for the Geological Survey of India and during part of this period he was also director of the Geological Museum in Calcutta.

Following a devastating earthquake centred on Assam in June 1897, Oldham showed that the deep shock waves from a quake divide into two different types – the P or primary waves, which travel quickest, and the S or secondary waves, which arrive at seismographs later and vibrate at right angles to the direction of travel.

In 1903, Oldham returned to the UK, partly due to poor health. In 1906, he published a report analysing P and S wave arrivals from a group of earthquakes, from which he concluded that the Earth must have a dense central core with a radius less than 0.4 times the Earth's radius. The seismological data he used to demonstrate this would today be considered inadequate, but his work spurred more vigorous research to clarify the Earth's interior structure.

DAVID HILBERT
23 January 1862 – 14 February 1943

*German mathematician who set the maths research
agenda for the 20th century*

'Who of us would not be glad to lift the veil behind
which the future lies hidden; to cast a glance at the
next advances of our science and at the secrets of its
development during future centuries? What particular
goals will there be toward which the leading mathematical
spirits of coming generations will strive?'

From Hilbert's 1901 lecture 'Mathematical Problems'

Born in Königsberg, East Prussia (now
Kaliningrad, Russia), Hilbert received his doctorate
from the university in his home town in 1885. He
became professor there from 1886 until 1895, when
he was appointed chairman of mathematics at the
University of Göttingen, where he remained for the
rest of his career.

One of Hilbert's many great contributions to
mathematics was to give Euclidean geometry a rigorous axiomatic footing.
Euclid had either knowingly or unknowingly made more assumptions
in his *Elements* than he stated. After systematically studying Euclid's
theorems, Hilbert proposed and explained the significance of 21 axioms –
logical statements necessarily assumed to be true – in Euclidean geometry,
and he published the results in *The Foundations of Geometry* (1899).

Hilbert also introduced the mathematical concept of the 'Hilbert
space', a space with any finite or infinite number of dimensions. This
proved a vital tool in the later development of many theories in physics
including quantum mechanics. In 1900, he defined an influential list of
23 important unsolved problems in maths. These became the backbone
of 20th-century research in maths and most of them have been at least
partially solved.

PIERRE CURIE
15 May 1859 – 19 April 1906

*Polish–French husband and wife team famous for their
pioneering studies of radioactivity*

'It can even be thought that radium could become very
dangerous in criminal hands, and here the question can be
raised whether mankind benefits from knowing the secrets
of Nature, whether it is ready to profit from it or whether
this knowledge will not be harmful for it. The example of the
discoveries of Nobel is characteristic, as powerful explosives
have enabled man to do wonderful work. They are also a
terrible means of destruction in the hands of great criminals
who are leading the peoples towards war. I am one of those
who believe with Nobel that mankind will derive more good
than harm from the new discoveries.'

From Pierre Curie's Nobel lecture (6 June 1905)

 **Marie Curie was born in Warsaw.** Her
father encouraged her interest in science and in 1891
she moved to Paris to study physics and maths at the
Sorbonne. In 1894, she met Paris-born Pierre Curie,
who was appointed professor at the Sorbonne in
1895, the same year that he and Marie married. Marie
earned her doctorate in 1903 under the supervision
of **Becquerel**.

After Pierre was killed in a road accident involving a horse-drawn
vehicle in 1906, Marie succeeded him as professor of physics at the
Sorbonne, becoming the first woman professor at the university. She was
later appointed director of the Curie Laboratory in the Radium Institute of
the University of Paris, founded in 1914.

Pierre's early work focused on crystallography and magnetism,
in which he showed that the magnetic properties of a material change

at a certain critical temperature, now known as the Curie point. But they are both most famous for their work on radioactivity. Inspired by the discovery of radioactivity by Becquerel, Marie Curie meticulously studied the radiation of all known compounds containing radioactive elements and showed that it's possible to exactly measure the strength of the radiation from uranium or thorium, which is proportional to the amount of uranium or thorium in the compound, regardless of what the compound is.

Profoundly, Marie Curie showed that the radiation goes deeper than chemistry in that it doesn't depend on the arrangement of the atoms in a molecule – the process must originate in the interior of the atom itself. Noticing that pitchblende (a uranium ore) had stronger radiation than uranium alone, Marie Curie suspected it must contain an unknown element with more powerful radiation than uranium.

She developed new methods for separating the radioactive elements in pitchblende with Pierre and together they isolated two new elements in 1898: radium and polonium, which they named after Marie's home country, Poland. Separating large amounts to study was an arduous task. They managed to separate a tenth of a gram of radium chloride from a tonne of pitchblende by 1902, and after Pierre's death, Marie isolated pure radium metal by 1910.

For their studies of radioactivity, Marie and Pierre Curie shared half of the 1903 Nobel prize for physics; the other half went to Becquerel. In 1911, Marie Curie won the Nobel prize for chemistry for the discovery and study of radium and polonium. She remains the only woman in history to have won a Nobel prize twice (and her daughter **Irène Joliot-Curie** went on to win a Nobel award in 1935). Marie Curie died of aplastic anaemia, a blood disease that often results from exposure to large amounts of radiation.

*German–American scientist considered the founder of
modern anthropology*

'The historical development of the work of anthropologists seems to
single out clearly a domain of knowledge that heretofore has not been
treated by any other science. It is the biological history of mankind
in all its varieties; linguistics applied to people without written
languages; the ethnology of people without historic records; and
prehistoric archaeology.'

From Boas's article *The History of Anthropology* (1904)

Born in Minden, Westphalia, Boas studied
at the universities of Heidelberg, Bonn and Kiel,
where he earned a PhD in physics in 1881. In 1883, he
travelled to Baffin Island to research the impact of the
physical environment on Inuit migrations and in 1885
he returned to Germany to work with **Virchow** at the
Royal Ethnological Museum in Berlin. In 1896, Boas
became a curator at the American Museum of Natural
History in New York and a lecturer at Columbia University. Three years
later, he became the first professor of anthropology at Columbia.

Boas robustly discredited pseudoscientific theories of racial superiority
in the early 20th century. His book *The Mind of Primitive Man* (1911)
demonstrated that there was no such thing as a 'pure' race. In a study of
US immigrants and their descendants, he showed that environment plays
a significant role in determining physical attributes, like head size, which
had often been used to demarcate racial differences. He also exhibited
skulls from various communities to illustrate the absurdness of linking
superiority to brain size.

He promoted application of the scientific method in anthropology,
urging his peers to avoid value judgements, emotional prejudice and
sweeping generalizations. He also used statistical studies to demonstrate
how much human variation is context-dependent, arguing that this
dependency made many generalizations, including popular theories of
social evolution, completely unscientific.

ARTHUR JOHN EVANS
8 July 1851 – 11 July 1941

British archaeologist who uncovered the
Palace of Knossos in Crete

'Knossos lies in what, as far as human records go back, has always
been a great seismic centre. Earthquake after earthquake laid the
Palace low; always it rose again from its ruins more magnificent
until that final disaster from which there was no recovery.'

From J.D.S. Pendlebury and Arthur Evans's *Handbook to the Palace of Minos at Knossos
with its Dependencies* (1933)

Evans studied history at the University of Oxford from 1870 to
1874. During this period he also travelled widely making notes about all
the places, cultures and historical artefacts he came across. After that he
briefly studied at the University of Göttingen in Germany, and from 1884
to 1908, he was keeper of the Ashmolean Museum at Oxford.

Evans first visited Crete in 1894 to study unknown scripts on Cretan
'seals', tiny stones that ancient Mediterranean civilizations carved with
inscriptions. In this work, he identified the Minoan hieroglyphs and
pre-alphabetic scripts now known as Linear A and B. In 1900, he began
work at Knossos, a site discovered in 1878 near Crete's capital Heraklion.
Evans bought the site with inherited money and employed local labourers
to excavate it.

By 1903, excavations had revealed around 1,300 interlocking rooms,
with a theatre, workrooms, living spaces and store rooms. Intricate
frescoes decorated many of the walls. The site was built from around
7000 BC and probably served as a ceremonial and administrative centre
of the flourishing Minoan civilization and culture, populated by up to
about 100,000 people. Evans named it the 'Palace of Minos' believing,
controversially, that it was the palace of the legendary King Minos from
Greek mythology. He was knighted in 1911.

MAX KARL ERNST LUDWIG PLANCK
23 April 1858 – 4 October 1947

German physicist who pioneered early quantum theories

'How do we discover the individual laws of physics, and what is their nature? It should be remarked, to begin with, that we have no right to assume that any physical law exists, or if they have existed up to now, that they will continue to exist in a similar manner in the future. It is perfectly conceivable that one fine day, Nature should cause an unexpected event to occur which would baffle us all; and if this were to happen, we would be powerless to make any objection, even if the result would be that, in spite of our endeavours, we should fail to introduce order into the resulting confusion. In such an event, the only course open to science would be to declare itself bankrupt.'

From Planck's *The Universe in the Light of Modern Physics*

Planck studied physics at the universities of Munich and Berlin, where his teachers included **Kirchhoff** and **Helmholtz**, and he earned his doctorate at Munich in 1879. He taught at Munich from 1880 to 1885 then became associate professor of theoretical physics at Kiel University. In 1889, he succeeded Kirchhoff as professor at Berlin University on Helmholtz's recommendation. He remained there until his retirement in the late 1920s.

In his early career, Planck's work led to important advances in thermodynamics and radiation. Experiments had measured the radiation emitted by a black body (an idealized physical body that absorbs all incident electromagnetic radiation in thermal equilibrium) and shown that the distribution of wavelengths didn't match the predictions of classical physics. The classical picture was that the amount of radiation of a given frequency should rise with the number of the radiator's natural modes in that range, in other words the number of full or half wavelengths of the radiation that a cavity could sustain.

But the predicted continual increase in radiated energy with frequency – the so-called 'ultraviolet catastrophe' – clearly didn't happen in reality. Planck applied quantum theory to the problem, suggesting the energy emitted by a resonator can only take on discrete values or quanta. He was able to derive a formula that correctly matched the distribution of wavelengths from a black body of a given temperature. The equation involved a universal constant h, now called Planck's constant.

Planck's formula correctly describes how a black body emits a kind of skewed bell curve of wavelengths, with the peak wavelength depending on the body's temperature. Extremely hot bodies peak in the X-ray waveband, for instance, while cooler ones peak in the infrared. It explains how as a cool body progressively heats up, it glows red hot, then white, then blue when seen by the human eye (which doesn't register the full breadth of the bell curve because we can't see the portions in infrared or ultraviolet and beyond).

However, the theory took time to gain general acceptance. Planck himself was not completely certain that his equation wasn't just a mathematical artefact, something that looked roughly right but didn't rest on any real mechanism in nature. However, it became generally accepted around 1913 when **Bohr** applied quantum theory to the structure of the atom and correctly predicted the frequencies of the spectra for the hydrogen atom. And in 1924, **Bose** reported that he had derived Planck's quantum radiation law using **Boltzmann**'s statistics.

In later life, Planck researched a range of topics in physics and chemistry, including optics, thermodynamics and statistical mechanics. He was also the first prominent physicist to champion **Einstein**'s special theory of relativity, published in 1905. He encouraged Einstein and **von Laue** to come to work at Berlin and when Planck retired, **Schrödinger** was appointed as his successor. For many years, Berlin was a thriving centre for theoretical physics until the rise of the Nazis during the 1930s.

Planck's life was overshadowed by personal tragedy. Three of his children died young, leaving him with two sons. During the Second World War his home was destroyed during a bombing raid and one of his remaining sons, Erwin Planck, was arrested in July 1944 for his role in an unsuccessful attempt to assassinate Hitler. He was sentenced to death and executed at Plötzensee Prison in Berlin in January 1945. Planck himself died less than three years later.

The European Space Agency's Planck spacecraft mission, launched in 2009 to study the cosmic microwave background (the relic radiation of the Big Bang) is named in his honour, as is the Max Planck Society, which operates many prestigious scientific research institutes.

KONSTANTIN EDUARDOVICH TSIOLKOVSKY
17 September 1857 – 19 September 1935

Soviet scientist considered one of the founding fathers of rocketry and astronautics

'Man will not always stay on Earth; the pursuit of light and space will lead him to penetrate the bounds of the atmosphere, timidly at first, but in the end to conquer the whole of solar space'

The epitaph Tsiolkovsky wrote for himself

Tsiolkovsky grew up in a village in Ryazan Province, south of Moscow, and developed a childhood illness that left him almost completely deaf. From the age of about 14, he taught himself from textbooks and from 16 he lived in Moscow, studying physics, maths and chemistry from library books. He also read about spaceflight in science fiction by Jules Verne, which had a remarkable influence on many early spaceflight pioneers.

Returning to his home region, Tsiolkovsky was appointed as a maths teacher in a local school and in his spare time carried out scientific experiments. He described the kinetic properties of gases, and speculated on the effects that the vacuum and weightlessness might have on future space travellers. During his lifetime he published almost 100 works relating to space travel, and they included designs for rocket steering systems, multistage rockets, space stations and life support systems. He also anticipated interplanetary travel and built a wind tunnel to experiment with aerodynamics, determining the drag coefficients for objects like cylinders and cones.

Tsiolkovsky also formulated important mathematical relationships in rocketry. His major work *The Exploration of Cosmic Space by Means of Reaction Devices* (1903) described what's now known as the Tsiolkovsky equation, which relates the delta-v of a rocket (the maximum change in its speed if no other external forces act) to the effective exhaust velocity and the rocket's initial and final mass. He later determined other important parameters, such as the escape velocities necessary to exit gravity fields. Along with **Oberth** and **Goddard**, who came to many similar conclusions independently, he is considered one of the founders of the space age that later followed.

KARL LANDSTEINER
14 June 1868 – 26 June 1943

*Austrian-born American biologist who recognized the main
blood groups and co-discovered the polio virus*

'It soon became clear, however, that the reactions follow a
pattern, which is valid for the blood of all humans, and that
the peculiarities discovered are just as characteristic of the
individual as are the serological features peculiar to an animal
species. Basically, in fact, there are four different types of
human blood, the so-called blood groups.'

From Landsteiner's Nobel lecture (11 December 1930)

Born in Vienna, Landsteiner studied medicine
at the University of Vienna, graduating in 1891.
For the following five years, he worked in chemical
research laboratories in Zurich, Würzburg (with
Fischer) and Munich. After further medical studies
and research he became professor at Vienna
University in 1911. Following a short spell working
at a hospital in The Hague, he emigrated to the
US in 1922 and held posts at the Rockefeller Institute in New York until
his death.

At the start of the 20th century, it was unclear why blood transfusions
sometimes caused rapid death from conditions such as shock and
jaundice. In 1901, Landsteiner showed that this occurs when the blood
from two people agglutinates on contact. However, it did not happen if
both people have the same blood groups, which he identified as A, B and
C (later changed to O). With the group AB added later, this blood typing
system led to the first safe blood transfusions.

For his work on blood groups, which also led to techniques for forensic
analysis and paternity testing, Landsteiner won the 1930 Nobel prize for
physiology or medicine. With Austrian physician Erwin Popper, he was
also first to isolate the virus that causes poliomyelitis.

WILLIAM MADDOCK BAYLISS
2 May 1860 – 27 August 1924

English physiologist who co-discovered secretin, the first known hormone

'There must never be the least hesitation in giving up a position the moment it is shown to be untenable. It is not going too far to say that the greatness of a scientific investigator does not rest on the fact of his having never made a mistake, but rather on his readiness to admit that he has done so, whenever the contrary evidence is cogent enough.'

From Bayliss's *Principles of General Physiology* (1915)

Bayliss studied medicine at University College London (UCL). He then moved to Wadham College at Oxford University to study physiology, graduating in 1888, before returning to UCL, where he worked for the rest of his life, becoming professor of physiology in 1912.

Research had already shown that the pancreas produces digestive juices in response to increased acidity in the duodenum and jejunum, two sections of the small intestine. **Pavlov** had suggested that the nervous system controls pancreatic secretions. To test this, Bayliss and his colleague Ernest Henry Starling severed the duodenal and jejunal nerves in anaesthetized dogs then introduced acid into the duodenum and jejunum. They showed that the pancreas still produced digestive juices, triggered by a chemical signalling process via the bloodstream. The intestinal lining secreted a chemical messenger they named 'secretin' – the first known hormone – into the blood.

Bayliss was drawn into political controversy by his vivisection experiments in the 'Brown Dog Affair'. Swedish activists alleged that he performed illegal dissection on a terrier in 1903. The procedure was condemned as cruel by the National Anti-Vivisection Society and led to riots in London. Antivivisectionist Stephen Coleridge accused Bayliss of breaking the law, but Bayliss sued for libel and was awarded damages of £2,000. He was knighted in 1922.

BERTRAM BORDEN BOLTWOOD
27 July 1870 – 15 August 1927

American chemist and physicist who made pioneering studies of radioactive decay

'The tragic death on 14 August 1927 of Bertram B. Boltwood, professor of radiochemistry in Yale University, removes an outstanding scientific personality who played an important part in the rapid expansion of our knowledge of radioactive transformations in the early days of radioactivity'

From Ernest Rutherford's obituary of Boltwood (*Nature*, 14 January 1928)

Born in Amherst, Massachusetts, Boltwood studied chemistry at Yale University in Connecticut from 1889, then spent two years at the University of Munich, returning to Yale as a researcher in 1894. After working as a private consultant from 1900 to 1906, he held several professorships at Yale but suffered a breakdown in 1922 from which he never fully recovered. He committed suicide in 1927.

In 1903, **Rutherford** and Frederick Soddy had suggested that radium might be a radioactive decay product of uranium, but attempts to prove this had failed. Boltwood suggested that some intermediary decay product formed first and began a search for it in 1904. In meticulous experiments, he found the intermediary he was looking for and believing it to be a new element, named it 'ionium'. This intermediary element later turned out to in fact be a type of thorium; Boltwood was close to the idea that atoms come in different varieties (now known to be isotopes, which have different numbers of neutrons in their atomic nuclei).

Boltwood also correctly proposed after mineral experiments that lead is the final, non-radioactive decay product in the uranium–radium decay chain. In 1907, he was first to suggest this would make radioactive dating of the Earth's crust possible; the older the rocks, the more uranium would have decayed to lead. Measuring these ratios later became a standard method for dating ancient rocks.

ALBERT EINSTEIN
14 March 1879 – 18 April 1955

German–Swiss–American physicist who radically altered our picture of the laws of nature

'Concern for man himself and his fate must always form the chief interest of all technical endeavours ... in order that the creations of our minds shall be a blessing and not a curse to mankind. Never forget this in the midst of your diagrams and equations'

From Einstein's address to students at the California Institute of Technology
(*The New York Times*, 17 February 1931)

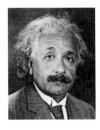

Einstein was born to non-observant Jewish parents in Ulm, Württemberg, and grew up in Munich. In 1896, he enrolled at the Swiss Federal Polytechnic School in Zurich to train as a physics and maths teacher and in 1901, he earned his diploma. Unable to secure a teaching post, he became a technical assistant at the Swiss Patent Office. During this time he published ground-breaking papers on theoretical physics and he was awarded a PhD by the University of Zurich in 1905.

In 1909, Einstein was appointed professor in Zurich and from 1914 he was director of the Kaiser Wilhelm Physical Institute and professor at the University of Berlin. In 1933, he emigrated to the US, where he was appointed professor at Princeton University in New Jersey. He became a naturalized US citizen in 1940. Following the Second World War, he was offered the post of second president of Israel, which he declined.

One of Einstein's first great achievements was to nail the case for the particle-like nature of light when he explained the 'photoelectric effect', showing that an individual photon of blue light has enough energy to dislodge a single electron from a metal, but red photons do not. This was a key discovery that led to the development of quantum mechanics, although Einstein famously held objections to quantum theory throughout his life. He disliked the probabilistic notion of indeterminism in quantum mechanics, famously declaring that 'I, at any rate, am convinced that

He [God] does not throw dice'. For his work on the photoelectric effect, he was awarded the 1921 Nobel prize for physics.

By that time, he had also published both his theories of relativity, which were still controversial at the time of his Nobel award. But it was this work that made him world famous. His special theory of relativity, published in 1905, was developed from two basic principles: the laws of physics must be the same for any observer moving at a constant velocity and the speed of light is always the same, regardless of the speed of the light source.

The theory abandoned the notion that it's possible to have a universal standard of time and space. Instead, the length of an object or time interval depends on who's measuring it. If a train was moving at close to the speed of light relative to an observer, for instance, the observer would perceive the train to be shorter than the passengers onboard would measure, while the observer would see a clock on the train run slow. It's not just an illusion – measurements show that unstable particles moving fast through the Earth's atmosphere decay much more slowly than they do at rest in a laboratory.

Einstein generalized his relativity theory to consider gravitational effects by 1915. Unlike Newtonian gravity, general relativity views gravity as a natural upshot of the geometry of curved space and ditches the notion that gravity is 'action at a distance'. Large masses like planets move in response to the curvature of space–time, distorted by mass itself. Matter tells space how to curve; curved space tells matter how to move.

General relativity makes some specific predictions, for instance quantifying the way the gravity of the Sun should deflect light from a background star. **Eddington** organized expeditions to observe this during a total eclipse of the Sun in May 1919 and it confirmed Einstein's predictions.

When Einstein rejected the honour of becoming second president of Israel, he wrote in his response: 'All my life I have dealt with objective matters, hence I lack both the natural aptitude and the experience to deal properly with people and to exercise official function. I am the more distressed over these circumstances because my relationship with the Jewish people became my strongest human tie once I achieved complete clarity about our precarious position among the nations of the world.'

JOHN SCOTT HALDANE
3 May 1860 – 15 March 1936

Scottish physiologist famous for intrepid self-experimenting

'All around us we see disease and death – facts which,
in themselves, we seem unable to interpret from a
psychological standpoint. Disease, crime, death and birth
are matters so familiar to us that the deep mystery which
surrounds them is scarcely realized ... We are accustomed
to think that since scientific knowledge has cleared up so
much that had previously been mysterious, our experience
must be much more intelligible to us than it seemed to our
forefathers. In so thinking, however, we forget that each
scientific advance seems only to throw into clearer relief
the mystery which remains.'

From Haldane's Gifford lectures 'The Sciences and Philosophy' (1926 – 1928)

Haldane graduated in medicine at Edinburgh University in 1884. After a brief research post at the University of Dundee, he moved to Oxford University in 1887, and from 1913 he was director of the Mining Research Laboratory at Bentley Colliery near Doncaster, which moved in 1921 to Birmingham University.

Haldane's work focused on applying physiological research to practical industrial and social problems. Struck by the high death rates of miners from carbon monoxide poisoning, he recommended the simple precaution of taking mice into the mine as an 'alarm' system, their higher metabolic rate making them succumb to the gas more quickly. Haldane was a famous self-experimenter, locking himself in sealed chambers and breathing toxic gases to measure their physiological effects. He often carried out these experiments with his son **J.B.S. Haldane.**

He also introduced guidelines for safe decompression for deep-sea divers. In work on respiration, he showed that an excess of carbon dioxide in arterial blood, rather than a lack of oxygen, triggers the nervous system's respiratory reflex.

NETTIE MARIA STEVENS
7 July 1861 – 4 May 1912

American geneticist who recognized the chromosomal basis of sex determination

'Since the somatic cells of the female contain 20 large chromosomes, while those of the male contain 19 large ones and 1 small one, this seems to be a clear case of sex determination, not by an accessory chromosome, but by a definite difference in the character of the elements of one pair of chromosomes'

Stevens describes her observations of sex determination in beetles
(*Studies in Spermatogenesis*, 1905)

Stevens was born in Cavendish, Vermont. She studied biology at Stanford University in California from 1896 and graduated with a master's degree in 1900. After that she worked with **Morgan** at Bryn Mawr College in Pennsylvania, where she earned a PhD in 1903. She later studied sex determination in worms and insects at the Carnegie Institution in Washington DC.

Stevens discovered that in some species, chromosomes differ markedly between the sexes. This was the first known indication that observable differences in chromosomes could be linked to observable differences in physical attributes, in this case, whether an individual is male or female. More specifically, she deduced the chromosomal basis of sex depended on the presence or absence of the Y chromosome.

Aged 50, Stevens died of breast cancer before she could occupy a research professorship that was created for her at Bryn Mawr College.

BERNARD BRUNHES
3 July 1867 – 10 May 1910

French physicist who discovered that the Earth's magnetic field has reversed in the past

'Four years before his death, Bernard Brunhes presented one of the most important findings in geomagnetism ... It was the first study to suggest the exciting possibility that reversed magnetization recorded by rocks was the result of a reversed geomagnetic field.'

From 'The Early History of Geomagnetic Field Reversals' by Maxwell Brown
(*The IRM Quarterly*, autumn 2010)

Brunhes studied physics at the École Normale Supérieure in Paris. He later taught at the University of Lille then at Dijon from 1897. In 1900, he moved to Clermont-Ferrand as professor of physics at the university there and head of the Puy-de-Dôme Observatory on an extinct volcano.

Brunhes became interested in the magnetization of rocks such as ancient lava flows, in which magnetic minerals line up with the Earth's magnetic field as the rocks solidify. In 1906, he showed that the field direction in an ancient basaltic flow in south-central France was opposite to that expected. He realized that this must be because the Earth's magnetic field flipped over in the past, with the magnetic north and south reversing.

This set off a debate that raged for more than 50 years. But by the early 1960s, it had become clear that there were indeed nine major magnetic reversals during the past 3.6 million years. Today it is accepted that Earth's magnetic field can exist in 'normal' and 'reversed' states, with a transition period of 2,000 to 10,000 years to change polarity. There's no consensus on why this happens. Bruhnes died aged just 42, probably from a brain haemorrhage.

CHARLES SCOTT SHERRINGTON
27 November 1857 – 4 March 1952

*English neurophysiologist who laid the foundations
of modern neurology*

'The brain is waking and with it the mind is returning. It is
as if the Milky Way entered upon some cosmic dance. Swiftly
the head-mass becomes an enchanted loom where millions
of flashing shuttles weave a dissolving pattern, always a
meaningful pattern though never an abiding one ... Strings of
flashing and travelling sparks engage the lengths of it. This
means that the body is up and rises to meet its waking day.'

From Sherrington's philosophical work *Man on His Nature* (1940)

Born in Islington, London, Sherrington
studied medicine at St Thomas's Hospital in London
and in 1879 went to Cambridge University to study
physiology. In the mid-1880s, he worked with
Virchow and **Koch** on bacteriology research, and from
1895, he was professor of physiology at the University
of Liverpool. From 1913 until his retirement in 1936,
he was professor of physiology at Oxford.

In experiments with primates, cats and dogs with selective brain
damage, Sherrington showed that nervous system reflexes are part of a
network of responses, as opposed to the contemporary notion that they
stemmed from isolated, independent 'reflex arcs'. He demonstrated that
when one set of muscles is stimulated, muscles opposing the action of
those muscles are also simultaneously inhibited.

His studies of the mammalian nervous system laid the foundations
of modern neurology and had practical benefits for brain surgery and
treatment of neurological disorders. Sherrington also coined the terms
'neuron' and 'synapse' to describe nerve cells and the structures that allow
neurons to pass electrical or chemical signals to adjacent cells. Knighted
in 1922, he won a half share of the 1932 Nobel prize for physiology or
medicine for his work on the nervous system.

HEIKE KAMERLINGH ONNES
21 September 1853 – 21 February 1926

Dutch physicist who pioneered cryogenics and discovered superconductivity

'It was a wonderful sight when the liquid, which looked almost unreal, was seen for the first time. It was not noticed when it flowed in. Its presence could not be confirmed until it had already filled up the vessel. Its surface stood sharply against the vessel like the edge of a knife. How happy I was to be able to show condensed helium to my distinguished friend Van der Waals, whose theory had guided me to the end of my work on the liquefaction of gases.'

From Kamerlingh Onnes's Nobel lecture (11 December 1913)

Kamerlingh Onnes was born in Groningen and enrolled at the local university in 1870. From 1871 to 1873 he studied under **Bunsen** and **Kirchhoff** in Heidelberg. After that he returned to Groningen, where he earned a doctorate in physics in 1879. In 1882, he was appointed professor of experimental physics and meteorology at Leiden University.

At Leiden, Kamerlingh Onnes set up a cryogenic laboratory to test theories about how fluids behave over a very wide range of temperatures. He devised innovative techniques for cooling gases, resulting in the first liquefied helium in 1908, and achieved a record low temperature by chilling helium down to just 0.9°C above absolute zero. For this work, he won the 1913 Nobel prize for physics.

Kamerlingh Onnes also demonstrated superconductivity for the first time when he chilled mercury to 4°C above absolute zero, showing that it then conducts electricity without any resistance. Superconductors have gone on to become vital components of medical scanners and levitating 'maglev' trains, which have achieved speeds of more than 580 km/h (360 mph).

HENRIETTA SWAN LEAVITT
4 July 1868 – 12 December 1921

American astronomer who discovered a revolutionary way to measure star distances

'Since the variables are probably at nearly the same distance from the Earth, their periods are apparently associated with their actual emission of light, as determined by their mass, density, and surface brightness'

Leavitt describes the link that allowed astronomers to measure star distances
(Harvard College Observatory Circular, 1912)

Born in Lancaster, Massachusetts, Leavitt studied at Radcliffe College in Cambridge, Massachusetts. From 1893, she worked at the Harvard College Observatory on a laborious analysis of the brightness of stars on photographic plates. She was struck by the large number of variable stars, ones that change in brightness with a regular period.

Examining variable stars (so-called Cepheid variables) in the Magellanic Clouds, small satellite galaxies of the Milky Way, Leavitt noticed an intriguing trend. In 1908, she published a report proposing that the brighter the star, the longer its period of variation. After further study, she confirmed this link by 1912 and showed that the relationship between period of variation and intrinsic brightness is surprisingly reliable.

That meant these stars could serve as a vital yardstick for measuring distances in the Universe. Soon after, the Danish astronomer Ejnar Hertzsprung calibrated the relationship by measuring the distance to several Cepheids in the Milky Way. Now, by measuring the period of variation of distant Cepheids, astronomers could estimate their distance accurately from their apparent brightness. This played a crucial role in **Hubble**'s work proving that many 'nebulae' in the night sky are in fact galaxies of stars in their own right, well beyond the Milky Way. Leavitt died of cancer aged 53.

THOMAS HUNT MORGAN
25 September 1866 – 4 December 1945

American biologist who clarified the role of chromosomes in heredity

'The most important contribution to medicine that genetics
has made is, in my opinion, intellectual ... The whole subject
of human heredity in the past (and even at the present time in
uninformed quarters) has been so vague and tainted by myths
and superstitions that a scientific understanding of the subject
is an achievement of the first order.'

From Morgan's Nobel lecture (4 June 1934)

Morgan was born in Lexington, Kentucky. He studied at the
University of Kentucky, where he earned his first degree in 1886, and
earned a PhD from Johns Hopkins University in Maryland in 1890. He
was associate professor of biology at Bryn Mawr College in Pennsylvania
from 1891 and in 1904 became professor at Columbia University in New
York. From 1928, he was professor at the California Institute of Technology
in Pasadena.

In breeding experiments with fruit flies, Morgan investigated the
transmission pattern of eye colour and other traits and showed that some
of them must be linked to the sex chromosome because the inheritance
patterns were influenced by whether offspring were male or female. Under
the microscope, he noted that while female flies had two identical-looking
X chromosomes, in males the X chromosome was paired with a different-
looking Y chromosome that is never present in the female.

From his experiments, Morgan formulated a chromosomal theory
of heredity which held that genes must reside on chromosomes, with
each gene having a specific location. He also showed that chromosomes
occasionally swapped portions ('crossover'). He wrote several classic works
on genetics, including *The Mechanism of Mendelian Heredity* (1915). For his
discoveries concerning the role of chromosomes in heredity, he won the
1933 Nobel prize for physiology or medicine.

FRITZ HABER
9 December 1868 – 29 January 1934

German chemist who synthesized ammonia for fertilizers and explosives

'The invention of this single man has enabled [Germany] not only to maintain an almost unlimited supply of explosives for all purposes, but to provide amply for the needs of agriculture ... It is a remarkable fact, and shows on what obscure and accidental incidents the fortunes of possibly the whole world may turn.'

Winston Churchill describes Haber's contribution (25 April 1918)

Born in Breslau (now Wroclaw, Poland), Haber studied chemistry under **Bunsen** at the University of Heidelberg from 1886. After further study at the University of Berlin and the Technical School at Charlottenburg, he worked for his father's chemical business before holding research posts in Karlsruhe. In 1911, he was appointed director of the Institute for Physical and Electrochemistry in Berlin-Dahlem.

Haber's most famous work involved the 'fixing' of nitrogen from air. Nitrogen makes up nearly four fifths of the atmosphere, yet production of fertilizers and explosives previously relied on surface nitrate deposits, often from remote countries like Chile. Haber invented the 'Haber process' that catalyses the reaction of atmospheric nitrogen and hydrogen at high pressure to create ammonia, which can then be oxidized to make nitrates and nitrites for fertilizers and explosives.

This soon led to industrial-scale ammonia production, enabling Germany to prolong the First World War after Allied forces blocked nitrate supplies. Haber also contributed to the development of chemical warfare, for which he was heavily criticized. But his invention of the Haber process still plays a vital role in maintaining food supplies worldwide. For his work, he won the 1918 Nobel prize for chemistry. As a Jew, he was forced to resign his post in 1933 and after a brief spell in England he died in Switzerland.

German medical scientist who developed pathogen-targeted drugs including a treatment for syphilis

'In order to pursue chemotherapy successfully we must look for substances which possess a high affinity and high lethal potency in relation to the parasites, but have a low toxicity in relation to the body, so that it becomes possible to kill the parasites without damaging the body to any great extent'

From a 1909 report by Ehrlich (*Readings in Pharmacology*, 1963)

Ehrlich was born in Strehlen, Silesia, then in Prussia and now in Poland. He studied at the universities of Breslau, Strassburg, Freiburg-im-Breisgau and Leipzig, where he earned his doctorate in 1878 for a dissertation on the theory and practice of staining animal tissues using dyes including **Perkin**'s aniline dyes. In the same year he became an assistant at the Berlin Medical Clinic where he continued his work on staining tissues with dyes.

In 1882, Ehrlich discovered an effective way to stain the tubercle bacillus discovered by **Koch**, who later appointed Ehrlich as his assistant. In 1896, he became director of a new institute at Steglitz in Berlin for serum research. After that he held several posts in Frankfurt.

In Berlin, Ehrlich collaborated with other scientists to develop a method of preparing and standardizing diphtheria antibodies from horses that had been infected with diphtheria and this led by 1892 to the development of an effective treatment for diphtheria, a dreaded disease with a high fatality rate, particularly in children.

In work on immunity, Ehrlich developed the 'side-chain theory' to explain how the body mounts a defence against disease. He argued that all cells have a variety of special receptors that he called side chains, and that they act like gatekeepers or locks for the cell. Each receptor had a unique structure and only substances with the right structure (like a key for the lock) could enter the cell.

The main function of receptors was to allow the cell to absorb nutrients for the cell. Unfortunately, many toxic substances also have the right 'key' to enter. Ehrlich proposed that when a toxin enters a cell, the body launches a defence strategy in which cells produce an excess of the relevant receptors that flood the body, mopping up and neutralizing free toxins by attaching to them. Although some of his ideas were wrong, Ehrlich's work laid important foundations for later research on immunity.

Ehrlich was convinced that discoveries from the fast-growing field of organic chemistry could allow the development of drugs that target a particular infectious agent. By identifying compounds that bind with particular chemical components, he reasoned that it should be possible to treat disease yet minimize side effects due to collateral damage to healthy cells. Inspired by his work on dyes, he reasoned that if a dye stained bacteria but did not stain ordinary cells, that must be because the dye is binding with the bacteria in a way that might prove fatal to the bugs alone. So that would constitute a 'magic bullet' that could seek out parasites in the body and destroy them without harming healthy cells.

He showed that one dye, trypan red, inhibited the trypanosomes that can cause sleeping sickness. To look for a better trypanosome killer, Ehrlich gathered and synthesized arsenic compounds with similar properties to trypan red. One of these chemicals, labelled 606 in his personal catalogue, was a failure against trypanosomes. But Ehrlich later asked his assistant to test its effect on the spirochete bacterium *Treponema pallidum* that causes the sexually transmitted disease syphilis. Untreated syphilis has a fatality rate of up to around 60 per cent.

Sure enough, it turned out that chemical 606 effectively killed the pathogen and Ehrlich's work led to the development of an effective treatment for the disease, Salvarsan, which was used from around 1910. It is considered the first example of a targeted drug therapy (or chemotherapy as Ehrlich called it, a term now normally associated with anticancer drugs). The drug also turned out to be successful at treating some other diseases including yaws, a tropical disease akin to syphilis, which a single injection of Salvarsan could cure.

In total, Ehrlich made and tested around 3,000 compounds in his search for drugs targeted against specific pathogens, including streptococci bacteria. For his work on immunity, he won a half share of the 1908 Nobel prize for physiology or medicine. In later life, he studied cancer and the immune system's response to it, and he became world-renowned in his own lifetime. He died during the First World War after suffering strokes.

ERNEST RUTHERFORD
(1ST BARON RUTHERFORD OF NELSON)
30 AUGUST 1871 – 19 OCTOBER 1937

British–New Zealand physicist who clarified the nature of radioactive decay and discovered the atomic nucleus

'I came into the room, which was half dark, and presently spotted Lord Kelvin in the audience and realized that I was in for trouble at the last part of my speech dealing with the age of the Earth, where my views conflicted with his. To my relief, Kelvin fell fast asleep, but as I came to the important point, I saw the old bird sit up, open an eye, and cock a baleful glance at me!'

From a speech by Rutherford at the Royal Institution in London, 1904

Rutherford studied maths and physical science at the University of New Zealand in Wellington, where he graduated with an MA in 1893 and a BSc in 1894. That same year, he was awarded an 1851 Exhibition Science Scholarship to study at Cambridge University's Cavendish Laboratory under **J.J. Thomson.**

In 1898, Rutherford moved to Canada as Macdonald Professor of Physics at McGill University in Montreal. He returned to England in 1907 to become professor at the Victoria University of Manchester (now Manchester University), and in 1919 he succeeded J.J. Thomson as Cavendish Professor of Experimental Physics at Cambridge.

While working in Cambridge with J.J. Thomson, Rutherford invented a detector for electromagnetic waves and investigated the behaviour of gases ionized by X-rays. In 1898, he reported measuring alpha and beta radiation from uranium, and having moved to Montreal, he went on to study 'emanations' from the element thorium and discovered a new inert gas, an isotope of radon later known as thoron.

At this time, the nature of radioactive decay was still unclear. Working with English radiochemist Frederick Soddy from 1900, Rutherford created the 'disintegration theory' of radioactivity that correctly proposed radioactivity is an atomic process, rather than some kind of molecular one. In studies of decay rates, he discovered the concept of the 'half life' –

a radioactive material always takes the same amount of time for half the sample to decay. He used this effect to carry out crude radioactive dating of a rock and concluded it was 40 million years old, fuelling debates about the age of the Earth, which at the time was wildly underestimated.

Rutherford also noticed that a third type of radiation from radium, discovered by a French chemist in 1900, was unusually penetrating and he named it gamma radiation (now known to be high-energy electromagnetic waves). For his work on radioactive elements, Rutherford won the 1908 Nobel prize for chemistry. But his most famous work was yet to come. Working with the German physicist Hans Geiger in Manchester, Rutherford developed a detector for single alpha particles (an early type of Geiger counter). With English physicist Ernest Marsden, he fired alpha particles at a thin gold foil. Most flew straight through the foil in a straight line, but a tiny number bounced off it at large angles. Rutherford soon realized these ones had happened to hit a tiny, positively charged nucleus at an atom's centre.

This led Rutherford to develop a model of the atom that was radically different to J.J. Thomson's 'plum pudding' model, in which negative electrons milled around in a smooth ball of positive charge. Now, it seemed that practically the whole mass of the atom and all its positive charge are concentrated in a tiny nugget at the atom's centre. Later developments by **Bohr**, **Heisenberg** and others gave this model rigorous foundations within the framework of quantum mechanics.

Rutherford also became the first person to deliberately induce radioactive decay. In 1917, he showed that the nuclei of some light elements such as nitrogen can be made to emit fast protons when hit by energetic alpha particles. The English physicist Patrick Blackett later showed in cloud-chamber experiments that in the process, nitrogen is transformed into an oxygen isotope.

In 1921, working with Bohr, Rutherford proposed the existence of the neutron to explain the stability of the atomic nucleus. The neutrons would create an attractive strong force to bind the nucleus together, overcoming the repulsive forces between the positively charged protons. His colleague **Chadwick** proved the existence of the neutron in 1932.

Rutherford was knighted in 1914 and raised to the peerage as Baron Rutherford of Nelson in 1931. His ashes were buried in Westminster Abbey, near the tombs of **Newton** and **Kelvin**.

ROBERT ANDREWS MILLIKAN
22 March 1868 – 19 December 1953

American physicist whose 'oil-drop experiment' revealed the charge on the electron

'There remained, however, some doubters, even among those of scientific credentials ... The most direct and unambiguous proof of the existence of the electron will probably be generally admitted to be found in an experiment which for convenience I will call the oil-drop experiment.'

From Millikan's Nobel lecture (23 May 1924)

Millikan initially studied classics at Oberlin College in Ohio from 1886 before earning a PhD in physics from Columbia University in New York in 1895. After a year studying in Europe, in 1896 he was invited by **Michelson** to become an assistant at the Ryerson Laboratory of the University of Chicago, where he was appointed professor in 1910. In 1921, he joined the California Institute of Technology and he remained there until his retirement in 1946.

Millikan made many major discoveries in the fields of electricity and optics. But he is best known for an experiment that determined the charge on the electron for the first time. He sprayed oil drops into a chamber with a vertical electric field, and each droplet picked up a slight charge of static electricity as it travelled through the air. By altering the voltage and measuring the motions of many droplets due to gravity and the electric field, Millikan showed that the charge can only assume fixed multiples of a small value, the charge of a single electron.

This removed any lingering doubts that the electron is a real entity that always has the same charge, and for this work, Millikan won the 1923 Nobel prize for physics. A prolific and philosophical author, he also wrote on the reconciliation of science and religion.

VICTOR FRANCIS HESS
24 June 1883 – 17 December 1964

Austrian–American physicist who discovered cosmic rays

'From a consideration of the immense volume of newly discovered facts in the field of physics, especially atomic physics, in recent years it might well appear to the layman that the main problems were already solved and that only more detailed work was necessary. This is far from the truth, as will be shown by one of the biggest and most important newly opened fields of research ... that of cosmic rays.'

From Hess's Nobel lecture (12 December 1936)

Hess was born at Waldstein Castle near Peggau in Steiermark, Austria, where his father was a forester. He was educated at Graz University, where he earned his doctorate in physics in 1910. From 1910 to 1920 he worked at the Institute of Radium Research of the Viennese Academy of Sciences and after that became professor at the universities of Graz and Innsbruck. He emigrated to the US with his Jewish wife in 1938 to escape Nazi persecution. There he was appointed professor at Fordham University in New York.

Hess is best known for work he carried out in 1911 to 1913. For years, scientists had been puzzled by high levels of ionizing radiation in the atmosphere. Hess designed high-precision instruments to measure this radiation and then undertook dangerous balloon flights to measure the radiation at altitudes of up to 5.3 km (3.3 miles).

He discovered that the level of ionizing radiation at 5 km (3.1 miles) is roughly twice that at sea level and correctly concluded that radiation comes from outer space, which was confirmed in 1925 by the American physicist **Millikan**, who gave the radiation the name 'cosmic rays'. Today they're known to be energetic particles accelerated through space by astronomical sources such as supernova explosions. For his work on cosmic rays, Hess won a half share of the 1936 Nobel prize for physics.

WILLIAM HENRY BRAGG
2 July 1862 – 10 March 1942

(WILLIAM) LAWRENCE BRAGG
31 March 1890 – 1 July 1971

British father-and-son team who pioneered X-ray crystallography to study the structure of materials

'The study of structure by means of a microscope is limited by the coarseness of the light which illuminates the object, for we can never hope to see details smaller than the wavelength of the light. By using X-rays with their very short wavelengths, this limit of minuteness has at one step been decreased ten thousand times ... We are actually looking into the interior of the molecule and the atom with this fine-grained form of light.'

From Lawrence Bragg's Nobel lecture (6 September 1922)

 **William Bragg (pictured) was born in Cumberland** and studied maths and physics at Cambridge University from 1881. In 1885, he was appointed professor of maths and physics at the University of Adelaide, South Australia, and from 1909 he held professorships at Leeds, University College London and the Royal Institution in London. In the early 1900s, Bragg researched the range of alpha particles emitted by radioactive elements and during the First World War, he developed acoustic techniques for locating enemy submarines.

His son Lawrence Bragg was born when William Bragg was living in Adelaide, where Lawrence studied maths at university, graduating in 1908. He came to England with his father in 1909 and studied at Cambridge University. In 1919, he was appointed professor at Manchester University, a post he held until 1937. He was Cavendish Professor of Experimental Physics at Cambridge from 1938 to 1953.

After moving to England, the Braggs heard about **von Laue**'s work. Von Laue had shown that crystals act as diffraction grating for X-rays, the molecules making the X-rays bend and then constructively and destructively interfere in characteristic, regular patterns. (A related effect explains why the grooves in CDs reflect optical light in colourful patterns.) Studying the effect, Lawrence Bragg developed a mathematical formula (now called the Bragg law) that describes the necessary conditions for X-ray diffraction in terms of the wavelength of incident X-rays and the spacing between the crystal's lattice planes.

The father-and-son team went on to show that by measuring diffraction patterns from crystals, they could determine crystal structures. They used the technique to measure the distances between atoms in crystals such as diamond, sulphur and copper. Their work also revealed that some crystals such as sodium chloride contain arrays of spaced-out ions rather than molecules. The Braggs also studied biological molecules, including proteins, using X-ray crystallography.

The technique went on to spur many scientific fields, revealing the lengths and types of chemical bonds and atomic-scale differences between materials such as minerals and alloys. It has also clarified the structure and function of important biochemicals from DNA to vitamins and drugs. Even today, it's still the chief method for determining the atomic structure of new materials.

For their work on X-ray crystallography, the Braggs shared the 1915 Nobel prize for physics. Lawrence Bragg was only 25 at the time and he has remained the youngest Nobel laureate on record. William and Lawrence Bragg were both later knighted, in 1920 and 1941 respectively, and both served as director of the Royal Institution late in life.

FREDERICK GOWLAND HOPKINS
20 June 1861 – 16 May 1947

English biochemist who recognized the essential role of dietary vitamins

'No one can deny that the recorded experience of voyagers and explorers in the 18th century, and particularly perhaps the records of the British Navy which deal with the incidence and cure of scurvy, would have directed thought towards our modern conception of vitamins, had the times been ripe. The knowledge concerning nutrition was then, however, entirely vague and the days of experiment in such matters had not yet come.'

From Hopkins's Nobel lecture (11 December 1929)

Hopkins was born in Eastbourne, Sussex. He studied chemistry at the Royal School of Mines in London and at University College London, and then became an assistant to a Home Office expert on poisoning. Later he opted to study medicine at Guy's Hospital where he graduated in 1894 and taught for a further four years. In 1898, he was invited to research chemical aspects of physiology at Cambridge University, where he was appointed professor in 1914.

Hopkins made diverse discoveries in biochemistry. For instance, he discovered xanthine oxidase, an enzyme widely distributed in tissues and milk that catalyses the oxidation of the purine bases xanthine and hypoxanthine to uric acid.

In 1912, he reported the discoveries for which he is best known. In animal experiments, he showed that animals fail to thrive and ultimately die if they're fed a diet of purely proteins, carbohydrates, fats, minerals and water. He recognized that healthy diets must include small quantities of other unidentified chemicals, which he called 'accessory food factors' (now called vitamins). For this work, Hopkins won a half share of the 1929 Nobel prize for physiology or medicine. He was knighted in 1925.

MAX THEODOR FELIX VON LAUE
9 October 1879 – 24 April 1960

German physicist who discovered X-ray diffraction by crystals

'I was suddenly struck by the obvious question of the behaviour of waves which are short by comparison with the lattice-constants of the space lattice. And it was at that point that my intuition for optics suddenly gave me the answer: lattice spectra would have to ensue ... The ratio of wavelengths and lattice constants was extremely favourable if X-rays were to be transmitted through a crystal.'

From von Laue's Nobel lecture (12 November 1915)

Born in Pfaffendorf, near Koblenz, von Laue studied maths, physics and chemistry at the University of Strassburg. After further studies at the University of Göttingen and the University of Munich, he went to the University of Berlin in 1902 to work for **Planck**. He earned his doctorate there in 1903. After several research and lecturing posts, he became professor of physics at Berlin University from 1919 until 1943.

Von Laue's key discovery, published in 1912, was X-ray diffraction by crystals. At the time, it was unclear whether X-rays were particles or waves, but von Laue believed they were electromagnetic waves, like visible light. He recognized that the spacing of atoms in a crystal was larger, but not much larger, than the probable X-ray wavelengths, and therefore that X-rays should bend and interfere when passing through a crystal to create an interference pattern.

He mathematically modelled this and his experimental confirmation of X-ray diffraction by crystals firmly established X-rays as electromagnetic in nature. For this work, von Laue won the 1914 Nobel prize for physics. He died aged 80 after suffering injuries in a car crash.

ALFRED LOTHAR WEGENER
1 November 1880 – winter 1930

German scientist who introduced the idea of continental drift

'The forces which displace continents are the same as those which produce great fold-mountain ranges. Continental drift, faults and compressions, earthquakes, volcanicity, transgression cycles and polar wandering are undoubtedly connected on a grand scale.'

From Wegener's *The Origin of Continents and Oceans* (1966 translation)

Wegener was born in Berlin and studied physics, astronomy and meteorology at universities in Heidelberg, Innsbruck and Berlin. In 1906, he joined a meteorological expedition to Greenland, and later worked at the universities of Marburg and Graz, Austria.

Wegener is best known for proposing the idea of continental drift around 1912. He was struck by the similarity of the east coast of South America and the west coast of Africa, and noted that Antarctica, Australia, India and Madagascar could fit jigsaw-like to the tip of southern Africa. He became convinced that the continents had once been joined together in a single landmass (today known as Pangaea) that broke up, the continents gradually drifting apart over time.

Wegener backed up his idea by demonstrating similarities between rocks and fossils, especially fossilized plants, from both sides of the Atlantic Ocean. He also speculated that magma rises and solidifies at mid-ocean ridges where sea-floor spreading occurs and published his theories in *Die Entstehung der Kontinente und Ozeane* (*The Origin of Continents and Oceans*) in 1915. Opponents argued that the oceanic crust was too tough for the continents to move through and the theory was not generally accepted until the 1950s.

Wegener died on an expedition to Greenland to monitor Arctic weather during the winter of 1930. On 12 May 1931 his body was discovered, carefully buried by his expedition partner, who was never seen again.

ROBERT HUTCHINGS GODDARD
5 October 1882 – 10 August 1945

American physicist who is considered the father of modern rocket propulsion

'It is difficult to say what is impossible, for the dream of yesterday is the hope of today and the reality of tomorrow'

From Goddard's high school graduation oration (1904)

Goddard developed an interest in science and flight at an early age and studied physics at Worcester Polytechnic Institute, where he graduated with a BSc in 1908. He earned a PhD from the town's Clark University in 1911 and then accepted a research fellowship at Princeton University in New Jersey.

In 1914, Goddard was granted patents for the liquid-fuel rocket concept (proposed independently by **Oberth** and **Tsiolkovsky**) and for multi-stage rockets using solid fuel. He wrote in a 1920 report that it should be possible to send a rocket to the Moon with an explosive flash device to signal its arrival, an idea that provoked extensive ridicule in the press. In 1926, Goddard tested the first liquid-fuel rocket in Auburn, Massachusetts, and three years later he launched the first scientific payload, including a barometer and camera. By 1941, his team had launched 34 rockets to altitudes of up to 2.6 km (1.6 miles).

Goddard's work anticipated much of the technologies employed in **von Braun**'s German V-2 missiles, including gyroscopic control and power-driven fuel pumps. During the Second World War, Goddard contributed to the development of practical jet-assisted takeoff and liquid-propellant rocket motors capable of variable thrust. Although he secured more than 200 rocketry patents during his lifetime, Goddard was not a household name until the dawn of the space age that was spurred on by his pioneering work. NASA's Goddard Space Flight Center in Greenbelt, Maryland, is named after him.

NIELS HENRIK DAVID BOHR
7 OCTOBER 1885 – 18 NOVEMBER 1962

Danish scientist who radically improved understanding of atomic structure and quantum mechanics

'After we obtained details as to the constitution of the atom, this difficulty became still more manifest; in fact, so long as we confine ourselves to the classical electrodynamic theory we cannot even understand why we obtain spectra consisting of sharp lines at all ... The electrons would claim a continuous radiation of energy from the atom, which would cease only when the electrons had fallen into the nucleus.'

Bohr describes in his Nobel lecture the paradox that led him to apply quantum theory to the atom (11 December 1922)

Born in Copenhagen, Bohr was the son of an eminent physiologist who awakened Bohr's interest in physics as a child. In 1903, he enrolled at Copenhagen University, where he earned a master's degree in physics in 1909 and a doctorate in 1911. In the same year, he visited Cambridge in England where he followed the experimental work of **J.J. Thomson**. In 1912, he worked at **Rutherford**'s lab in Manchester.

After holding various short posts, he then returned to Copenhagen University as professor in 1916. During the Nazi occupation of Denmark during the Second World War, Bohr learned of his imminent arrest, and the Danish resistance helped him and his wife escape by sea to Sweden. He spent the last two years of the war in England and America, where he became involved with the Manhattan Project to develop atomic weapons. After the war, he remained at Copenhagen University for the rest of his life.

Following Rutherford's discovery of the atomic nucleus, Bohr recognized that it threw up a paradox. **Maxwell** had shown that oscillating charged particles emit electromagnetic radiation. So if negatively charged electrons in an atom were orbiting a positively charged nucleus, they too should emit electromagnetic radiation and lose energy.

The electrons would then fall straight into the nucleus. Yet stable atoms clearly do exist – why?

Bohr resolved the problem by applying the newly established science of quantum mechanics to the atom. He assumed that there must be 'stationary' orbits for electrons in which they don't radiate away energy and assigned these orbits quantized units of angular momentum.

It seemed natural that the discrete wavelengths of light emitted by atoms corresponded to electrons falling down one or more levels, while absorption of the same wavelengths signalled electrons jumping up again. With this model, Bohr was able to correctly predict a series of wavelengths emitted by the hydrogen atom. For this work, he won the 1922 Nobel prize for physics.

Bohr also worked on radioactivity and his liquid-drop model of the nucleus provided an early theoretical description of nuclear fission, in which a nucleus splits. The model describes the behaviour of nuclear particles – neutrons and protons – as similar to the molecules in a drop of liquid. If given sufficient extra energy, for instance by absorbing an extra neutron, the spherical nucleus would distort into a dumbbell shape before splitting at the neck into two nearly equal fragments. This simple model described the average properties of nuclei fairly well.

In quantum mechanics, Bohr also conceived the principle of 'complementarity': that it's possible for a physical entity to have several contradictory properties. An example is light, which behaves either as a wave or as a stream of particles depending on the experiment that is observing it. Throughout his working life, he had many debates with **Einstein** about the baffling and paradoxical issues of quantum mechanics. In his later years, Bohr campaigned for the peaceful application of atomic energy and the resolution of political problems arising from atomic weapons.

KARL SCHWARZSCHILD
9 October 1873 – 11 May 1916

German physicist who solved Einstein's general relativity equations, highlighting the possibility of black holes

'As you see, the war treated me kindly enough, in spite of the heavy gunfire, to allow me to get away from it all and take this walk in the land of your ideas'

From a letter from Schwarzschild to Einstein (1916)

Schwarzschild became interested in astronomy at a young age and published papers on binary orbits by the age of just 16. He studied at the universities of Strasbourg and Munich, where he earned his PhD in 1896. He later worked at the Kuffner Observatory in Vienna and the University of Göttingen before becoming director of the Astrophysical Observatory in Potsdam in 1909.

Serving in the German army during the First World War, Schwarzschild solved the differential equations in Einstein's theory of general relativity to describe the gravitational field outside a massive object such as a star or planet (making simple assumptions about it being spherical, uncharged and not rotating). His solution predicted the possibility of black holes – masses so dense that nothing, not even light, can escape the strong gravity around them.

The boundary of such a black hole, the distance that marks the 'event horizon' or 'point of no return', is now known as the Schwarzschild radius. While still active on the Russian front, Schwarzschild contracted a rare skin disease from which he died aged 42.

PAUL LANGEVIN
23 January 1872 – 19 December 1946

French physicist who pioneered using sonar to detect submarines

'This remark provides the means for any among us who wants to devote two years of his life, to find out what the Earth will be in two hundred years ... It is sufficient that our traveller consents to be locked in a projectile that would be launched from Earth with a velocity sufficiently close to that of light ... Returned to Earth he has aged two years, then he leaves his ark and finds our world two hundred years older.'

From Langevin's relativity text *The Evolution of Space and Time* (1911)

Langevin was born in Paris and studied at the École Normale Supérieure there. After that, he studied at the University of Cambridge under **J.J. Thomson**. He then returned to Paris where he earned his PhD in 1902 while working with **Pierre Curie**. He became professor of physics at the Collège de France in 1904 and at the Sorbonne in 1909.

Langevin is best known for his work on ultrasonic devices. During the First World War, he developed technologies to detect enemy submarines by bouncing ultrasonic waves off them and detecting the reflections. His technology was perfected after the war, leading ultimately to powerful sonar (short for sound navigation and ranging devices) techniques.

In his work on magnetism, Langevin devised the modern explanation of why some materials are attractive in a magnetic field and others repulsive (paramagnetism and diamagnetism) in terms of the spins of electrons within atoms. He also promoted **Einstein**'s relativity theories in France. Langevin was an outspoken opponent of fascism and during the Second World War he was arrested by the Nazis and placed under house arrest. He escaped to Switzerland in 1944 but returned to Paris after the war.

MILUTIN MILANKOVITCH
28 May 1879 – 12 December 1958

Serbian geophysicist and engineer who explained long-term changes in the Earth's climate

'He talked about my calculations with such excitement that I got quite embarrassed. I crouched in my seat in the top row of the amphitheatre as much as I could so that a glance from Wegener would not reveal my presence in the auditorium.'

Milankovitch recalls **Wegener**'s enthusiasm for his theories explaining past ice ages

Milankovitch was born to a Serbian family in Dalj, Austria–Hungary (today in Croatia). He studied civil engineering in Vienna from 1896, graduating in 1902, and then returned there after a year of military service to earn a PhD in 1904. From 1905, he held a post as engineer for commercial construction companies, developing a strong reputation for projects on aqueducts, factories and bridges, and in 1909 he was appointed professor in the applied maths department at Belgrade University.

During the First World War, Milankovitch was held as a prisoner of war in Budapest, but allowed to continue his studies. At this time, he was interested in explaining geological evidence that the Earth had swung in and out of deep ice ages. He realized that this must be linked to the amount of radiation from the Sun reaching the Earth's surface, and demonstrated mathematically that our planet has a 21,000-year 'precession' (twisting round) of its spin axis and elliptical orbit, and a periodic 40,000-year cycle in which the tilt of the axis nods up and down.

Today these combined variations are known as the Milankovitch cycles, and studies of deep-sea sediment cores confirmed in the 1970s that they have played a major role in determining ancient climates on Earth.

ARTHUR STANLEY EDDINGTON
28 December 1882 – 22 November 1944

English astrophysicist who publicized general relativity and showed there's an upper limit to a star's brightness

'From his central position man can survey the grandest works of Nature with the astronomer, or the minutest works with the physicist. Tonight I ask you to look both ways. For the road to a knowledge of the stars leads through the atom; and important knowledge of the atom has been reached through the stars.'

From a lecture Eddington gave to the British Association in Oxford (August 1926)

Born in Kendal, England, Eddington studied at Owens College, Manchester (which later became part of the University of Manchester), and Cambridge University, where he earned his MA in 1905. From 1906, he was assistant to the Astronomer Royal at Greenwich until 1913, when he became Plumian Professor of Astronomy at Cambridge at the young age of 31.

A lifelong Quaker pacifist, Eddington claimed conscientious objector status when called up for military service during the First World War. Around this time, he received early reports about **Einstein**'s general theory of relativity. Eddington's excellent understanding of general relativity led him to become the chief proponent of the theory in Britain. To test the theory, he organized expeditions to observe a total solar eclipse in 1919 from Brazil and West Africa. This showed that the Sun bends light from background stars as Einstein's theory predicts.

In work on stellar structure, Eddington showed that more massive stars must also be more luminous, and he calculated that there must be a limit to how massive and luminous stars can become before extremely high radiation pressure makes them unstable. His popular and philosophical science writing made him a household name between the wars, and he was knighted in 1930.

GEORGE WASHINGTON CARVER
1864/1865 – 5 JANUARY 1943

*American agricultural scientist who revolutionized
farming practice in the southern US*

'I wanted to know the name of every stone and flower and
insect and bird and beast. I wanted to know where it got its
colour, where it got its life – but there was no one to tell me.
I do not know how I learned to read and write, but I did in
some way, thanks to the Carvers.'

Carver quoted in *George Washington Carver, Scientist and Symbol* by Linda McMurry (1981)

Carver was born to black slave parents
in Diamond Grove (now just Diamond), Missouri; his
exact birth date is unknown. His father died before
he was born, and following the kidnapping of his
mother, his master Moses Carver and his wife raised
George and his brother as their own children after the
abolition of slavery.

The Carvers encouraged George to learn to read
and to write. He studied art and piano at Simpson College in Indianola,
Iowa, from 1890. Noticing Carver's gift for painting plants, one of his
teachers encouraged him to study botany. In 1891, Carver became the first
black student at Iowa State Agricultural College in Ames and in 1896, he
was appointed head of agriculture at the Tuskegee Institute in Alabama,
where he taught for 47 years.

Carver's research resulted in the creation of 325 products from
peanuts, including dyes, plastics and nitroglycerin, as well as more than
100 products from sweet potatoes. By promoting good farming practice
and the use of products from crops other than cotton, he dramatically
improved the rural economy for poor farmers, black and white. By the
1930s, he was world famous for his agricultural experiments and hailed as
a tireless campaigner for racial equality.

FREDERICK GRANT BANTING
14 November 1891 – 21 February 1941

Canadian physiologist who discovered and isolated insulin to treat diabetes

'Insulin enables the severe diabetic to burn carbohydrate ...
With the relief of the symptoms of his disease ... the pessimistic,
melancholy diabetic becomes optimistic and cheerful.'

From Banting's Nobel lecture (15 September 1925)

Born in Alliston, Ontario, Banting went to
the University of Toronto to study divinity, but soon
switched to a medical course. During the First World
War he joined the Canadian Army Medical Corps and
was wounded in France at the Battle of Cambrai. In
1919, he was awarded the Military Cross for heroism
under fire. In 1921, he returned to the University of
Toronto as a lecturer in pharmacology.

By now, Banting had become interested in diabetes, then a fatal disease
in which dangerous levels of sugar accumulated in the blood. Research
had already linked the disease to the pancreas, which seemed to secrete
a hormone (insulin) that regulates sugar metabolism. An insufficient
supply of insulin caused diabetes, so presumably if the chemical could be
isolated and injected into patients, their diabetes could be treated.

But all attempts to isolate the digestive juices from the pancreas had
destroyed the insulin. Banting had read about a way to make pancreatic
tissue degenerate and reasoned that this might leave the insulin intact and
extractable. Working in John Macleod's lab with assistant Charles Best,
Banting isolated a coarse insulin extract from a dog's pancreas and used
it to keep a severely diabetic dog alive for 70 days. More refined extracts
led to successful human trials and since then millions of people have lived
relatively normal lives with diabetes.

Banting and Macleod shared the 1923 Nobel prize for physiology or
medicine. Banting was knighted in 1934 and died in a plane crash in
Newfoundland during the Second World War.

HOWARD CARTER
9 May 1874 – 2 March 1939

*English archaeologist and Egyptologist who discovered
the tomb of Tutankhamun*

'At the moment we did not even want to break the seal, for
a feeling of intrusion had descended heavily upon us ...
We felt that we were in the presence of the dead King and
must do him reverence.'

From *The Tomb of Tut-Ankh-Amen* by Howard Carter and A.C. Mace

During his childhood Carter developed a talent for illustration and an interest in ancient Egypt. Aged 17, he travelled to Egypt and was tasked by the Egypt Exploration Fund in London to help record texts from tombs in Middle Egypt. Other archaeological projects in Egypt kept him busy until 1899, when he was appointed chief inspector of antiquities for the Egyptian government.

In 1905, the Earl of Carnarvon visited Egypt and decided to finance archaeological work by Carter at Thebes (now Luxor) and at the Nile river delta. In 1912, Lord Carnarvon was granted permission to excavate in the Valley of the Kings, but the work was delayed by the outbreak of the First World War. Eventually, in November 1922, Carter discovered the tomb of Tutankhamun (the pharaoh who ruled Egypt between about 1333 BC and 1323 BC, and died aged about 19). It contained thousands of beautifully preserved treasures including the mummified pharaoh's now-iconic burial mask.

Exploring the tomb and documenting the objects inside took nearly a decade. It gave amazing insight into the royal burials, mummification and tombs of the New Kingdom's 18th Dynasty. After his work on the tomb, Carter retired from archaeology and became a collector of antiquities as well as an agent for collectors and museums.

ARTHUR HOLLY COMPTON
10 September 1892 – 15 March 1962

American physicist who discovered the Compton scattering effect

'The language of science is universal and is a powerful force
in bringing the peoples of the world closer together. We are all
acquainted with the sharp divisions which religions draw between
men. In science there are no such divisions: all peoples worship at
the shrine of truth.'

From Compton's Nobel banquet speech (10 December 1927)

Compton was born in Wooster, Ohio. He earned a science
degree from Wooster College in 1913 and a PhD from Princeton University
in New Jersey in 1916. After spending a year as a physics instructor at
the University of Minnesota, he became a research engineer with the
Westinghouse Lamp Company in Pittsburgh, Pennsylvania, until 1919,
when he studied at Cambridge University with **Rutherford**. From 1920, he
held professorships at Washington University in St Louis, Missouri, and
the University of Chicago.

Compton is famous for his work on X-ray scattering. In 1922, he
showed that the wavelength of X-rays increases (and hence their energy
decreases) when they are scattered by free electrons. This effect, now
called the Compton effect, was a clear illustration that electromagnetic
radiation such as X-rays behaves like particles as well as waves, colliding
with electrons to make them recoil.

For this discovery, Compton won a half share of the 1927 Nobel
prize for physics. His other research led to a substantial revision of the
value of the charge on the electron. During the Second World War, his
work contributed to the establishment of the first controlled uranium
fission reactors and the large reactors that produced plutonium for the
atomic bomb dropped on Nagasaki. NASA's highly successful Compton
Gamma-ray Observatory, which orbited Earth from 1991 to 2000, was
named in his honour.

SATYENDRA NATH BOSE
1 JANUARY 1894 – 4 FEBRUARY 1974

Indian mathematician and physicist who developed the quantum statistics of particles now called 'bosons'

'I have ventured to send you the accompanying article for your perusal and opinion. I am anxious to know what you think of it ... Though a complete stranger to you, I do not feel any hesitation in making such a request. Because we are all your pupils though profiting only by your teachings through your writings.'

From a letter from Bose to Einstein, who promoted Bose's work in Europe (4 June 1924)

Bose was educated at Presidency College in Calcutta before being appointed lecturer at the University of Calcutta in 1916. In 1921, he moved to Dhaka University. After achieving international renown for his quantum theories developed in the early 1920s, he travelled in Europe, working with **Marie Curie** in France and **Einstein** in Germany. He was appointed professor on his return to Dhaka in 1926.

Bose was inspired by Einstein's discovery that light waves consist of particles called photons. Using this concept, in 1924 he mathematically derived **Planck**'s radiation formula, which quantifies the electromagnetic radiation emitted by a body depending on its temperature. Planck had arrived at his formula largely through trial and error; Bose put the derivation on a sound mathematical foundation.

Journals in Europe refused to publish his work, but when Bose sent a copy to Einstein, Einstein recognized its merits and used his influence to have the paper published. Einstein generalized Bose's work to cover all particles, including some atoms, that like photons have integer values of a property called spin. These particles are now known as 'bosons'.

In the mid-1920s, Bose and Einstein also predicted the existence of an exotic state of matter, the Bose–Einstein condensate, which could form when bosons crash down into their lowest possible energy states. In 1995, scientists in Colorado produced the first Bose–Einstein condensate by chilling a cloud of rubidium atoms to almost absolute zero.

LOUIS DE BROGLIE
(LOUIS-VICTOR-PIERRE-RAYMOND, 7TH DUC DE BROGLIE)
15 AUGUST 1892 – 19 MARCH 1987

French physicist who pioneered quantum mechanics by describing wave properties of the electron

'The electron can no longer be conceived as a single, small granule of electricity; it must be associated with a wave and this wave is no myth; its wavelength can be measured and its interferences predicted'

From de Broglie's Nobel lecture (12 December 1929)

De Broglie was born in Dieppe and studied history before earning a science degree in 1913. During the First World War, he was conscripted for military service and stationed for the army's wireless section at the Eiffel Tower, where he studied technical problems in his spare time. After the war, he earned a doctorate in quantum theory from the University of Paris in 1924. He was appointed as professor of theoretical physics at the University of Paris in 1932.

Around the time that de Broglie was writing his thesis, scientists were grappling with some strange inconsistencies in the subatomic world. Light seemed to behave both as waves and as discrete particles depending on how you observed it, an effect called wave–particle duality. De Broglie's great insight was that particles show this duality as well, so that an electron behaves as a wave as well as a particle. Experiments confirmed this in 1927, showing that electron beams could create wave-like interference patterns.

De Broglie showed that this naturally explains **Bohr**'s model of the atom. Electrons can only occupy energy levels that correspond to a whole number of wavelengths, such as two, three or four, explaining why the electron energy levels are discretely quantized. It's similar to the way a vibrating violin string only sustains certain fixed standing waves, otherwise they would destructively interfere.

EDWIN POWELL HUBBLE
29 November 1889 – 28 September 1953

American astronomer who showed that other galaxies lie well beyond the Milky Way and found key evidence that the Universe is expanding

'You will be interested to hear that I have found a Cepheid variable in the Andromeda nebula ... the distance comes out something over 300,000 parsecs'

In a letter to astronomer Harlow Shapley, Hubble announces dramatic evidence that the Andromeda nebula is an amazingly distant galaxy of stars something like 1 million light years away (February 1924)

Hubble developed an early childhood fascination for science, partly from reading science fiction including Jules Verne's novels. He also excelled in sports and became a keen boxer. He studied maths and astronomy at the University of Chicago, graduating with his first degree in 1910. He then attended the University of Oxford on a Rhodes scholarship to study law, but didn't pursue this as a career and returned to the University of Chicago, where he earned a doctorate in astronomy in 1917.

After army service during the First World War, in 1919 Hubble was appointed as a staff astronomer at the Carnegie Institution's Mount Wilson Observatory in California, where he remained until his death. When he began work at the observatory, it was newly equipped with the 2.5 m (8.3 ft) Hooker Telescope, the largest telescope in the world at the time. This allowed Hubble to carry out an extensive survey of so-called Cepheid variable stars, which show periodic variations in brightness.

In 1908, **Leavitt** had shown that these stars have a fixed relationship between their intrinsic brightness and period of variation, so by measuring the period and apparent brightness of a Cepheid variable star, it is possible to determine how far away the star lies. Using observations from the early 1920s, Hubble was able to show that stars in several 'nebulae' are enormously distant from the Earth and lie in discrete galaxies well beyond our own Milky Way.

For the first time, astronomers had proof that separate 'islands' of stars exist throughout the Universe. Previously, the American astronomer Harlow Shapley and others had dramatically underestimated the scale of the Universe and concluded that many fuzzy nebulae – actually distant galaxies – lay inside the Milky Way itself. Even Hubble himself underestimated the distances to galaxies, which later had to be increased to accommodate more modern data.

Hubble's second claim to fame is that he established a relationship between a galaxy's distance and its redshift, the shifting of the characteristic wavelengths of light the elements inside the galaxies emit. The American astronomer Vesto Slipher had already noted that the spiral nebulae, later identified by Hubble as galaxies, had different levels of redshift. Hubble showed for the first time that the most distant galaxies have the highest redshifts. This reflects the fact that the Universe has expanded during the light's travel time, and the more distant the galaxy, the more the light waves have been stretched.

This later became key observational evidence to back the Big Bang theory. If on large scales, galaxies are receding from each other, then that suggested all matter was much closer together in the distant past and pointed to the origin of the Universe in a very hot state of unimaginably high density. By projecting backwards, it was possible to use Hubble's measurements to calculate the age of the Universe. His observations suggested a value of about 2 billion years, much smaller than the modern estimate of 13.7 billion years, but nonetheless his work transformed cosmology from philosophical speculation into a quantitative science for the first time.

However, Hubble himself was non-committal about the reason for the redshift–distance relationship, keeping his mind open to the idea that it doesn't reflect expansion of the Universe, but might signal 'a hitherto unrecognized principle of nature'. Hubble also introduced a galaxy classification system, which formed the basis of the one still used today. Some scientists regard Hubble as the greatest astronomer in terms of changing paradigms since the eras of **Galileo**, **Kepler** and **Newton**. The Hubble Space Telescope, placed in orbit by Space Shuttle astronauts in 1990, was named in his honour.

WOLFGANG ERNST PAULI
25 April 1900 – 15 December 1958

Austrian physicist who laid foundations for quantum theory and introduced the 'exclusion principle'

'I have done a terrible thing. I have postulated a particle that cannot be detected.'

Pauli's remark about predicting the existence of the neutrino (quoted in the Nobel lecture by Frederick Reines, 8 December 1995)

Born in Vienna, Wolfgang Pauli studied physics at the University of Munich and earned his doctorate in 1921. He then worked with **Born** at the University of Göttingen and with **Bohr** in Copenhagen before becoming a lecturer at the University of Hamburg (1923 – 1928). After that he became professor of theoretical physics at the Federal Institute of Technology in Zurich, which he returned to following a spell at Princeton University in New Jersey during the Second World War.

Pauli was recognized as an excellent physicist when he published a masterly exposition of the theory of relativity aged just 20. But he is most famous for introducing the 'Pauli exclusion principle' of quantum mechanics, which states that no two identical fermions (particles with half-integer spin, which include electrons, protons and neutrons) can occupy the same quantum state simultaneously. For this work, which underpins the structure and chemical behaviour of atoms, Pauli won the 1945 Nobel prize for physics.

He was also first to predict the existence of an extremely light particle that carries energy away in radioactive beta decay (later named the neutrino by **Fermi**). Pauli realized that these particles must carry no electric charge, and hence they would be extremely 'slippery' and difficult to detect. However, in 1970, scientists succeeded in detecting neutrinos from a nuclear reactor, confirming their existence. Neutrinos also pour out from nuclear reactions in the Sun; the vast majority that come towards the Earth simply fly straight through our planet.

JOHN LOGIE BAIRD
13 August 1888 – 14 June 1946

Scottish inventor and pioneer of television

'On looking at the screen of the televisor I saw rapidly moving dots and lines of orange light which gradually formed themselves into a definitely recognized face. This image varied from time to time in clarity, but movements could be clearly seen, and the image, when clear, was unmistakeable.'

W. Sutcliffe, staff chief engineer on Cunard's SS *Berengaria*, describes watching a transmission by Baird from London to the mid-Atlantic on 6 March 1928

Baird was born in Helensburgh, Scotland. He studied electrical engineering at the Glasgow and West of Scotland Technical College (later the University of Strathclyde) and Glasgow University. He didn't graduate, however, because his degree course was interrupted by the First World War. He worked as an engineer for the Clyde Valley electric power company, but was forced to give this up due to poor health, and after some unsuccessful business ventures he settled in Hastings in 1923.

In Hastings, Baird rented a workshop and in March 1925, he demonstrated his first TV apparatus built almost entirely from odds and ends including some darning needles and a pair of scissors. He transmitted silhouette television images at Selfridges department store and later that year, he successfully transmitted greyscale images. In 1926, he unveiled a television system that could broadcast live greyscale images with a scan rate of 12.5 pictures per second.

In 1927, Baird transmitted a long-distance TV signal between London and Glasgow (by now, AT&T Bell Labs had achieved a long-distance transmission between New York and Washington DC). The following year he made the world's first colour transmissions and the first transmission across the Atlantic. Many other inventors contributed to the development of television and Baird's mechanical transmission system soon became obsolete.

German-born physicist who made key advances in quantum mechanics

'The facts known up to the end of the 19th century seemed to indicate that the world was a perfect mechanism ... if its configuration were known at a given instant its future behaviour could be predicted with certainty ... But then new facts were discovered, in the realm of atoms as well as in the stellar Universe, facts which did not fit in the mechanistic frame.'

From Born's Nobel banquet speech (10 December 1954)

Born was born in Breslau (now Wrocław, Poland) and studied at the universities of Breslau, Heidelberg, Zurich and Göttingen. At Göttingen, he studied under **Hilbert** and **Schwarzschild** and graduated in 1907. Back in Breslau during 1908 to 1909, he studied relativity theory, which he lectured on at the University of Chicago from 1912.

Born joined the German armed forces during the First World War and then became professor at Göttingen in 1921, where he collaborated with luminaries such as **Pauli** and **Heisenberg**. Having Jewish ancestry, he was forced to emigrate in 1933 and taught at Cambridge University in the UK for three years before becoming professor of physics at Edinburgh University, where he remained until his retirement in 1953.

In the mid-1920s, Born developed a matrix technique to iron out some of the mathematical inconsistencies in the newly developed theory of quantum mechanics. It was already well known that particles sometimes behave like discrete objects and sometimes like waves. Born put forward the probability interpretation of this behaviour – the wave 'guides' the particle so that the probability of finding a particle at any one point is proportional to the square of the wave function's amplitude at this point. For his work on quantum mechanics, he won a half share of the 1954 Nobel prize for physics.

Hungarian–American mathematician who envisaged the design of modern computers

'Truth is much too complicated to allow anything
but approximations'

Von Neumann quoted in *Fractals, Chaos, Power Laws* by Manfred Schroeder (1991)

Born to a Jewish family in Budapest, von Neumann studied mathematics, physics and chemistry at German and Swiss universities before earning a PhD in maths at the University of Budapest in 1926. He taught in Berlin until 1930, when he was invited to Princeton University, New Jersey. With **Einstein**, he became one of the first faculty members at the Institute for Advanced Study in Princeton and he remained there until his death.

Von Neumann is generally viewed as one of the greatest mathematicians in modern history. In his early work in Germany, he developed the first rigorous mathematical framework for quantum mechanics. During the Second World War, he contributed to the Manhattan Project to develop nuclear fission weapons and later, working with **Teller** and others, he contributed to the design of the more powerful hydrogen (fusion) bomb. He also made important discoveries in pure mathematics, including geometry, set theory, probability and statistics.

But von Neumann is probably best known for his work on digital computers with modern architecture in which both programs and data are stored in the same memory. This led to the development of the first computers that could execute any kind of calculation without the need for hardware modification. His logical architecture dramatically increased the speed of future generations of computers.

A passionate critic of both the Nazis and left-wing extremism, von Neumann described his political ideology as 'violently anti-communist' during a Senate committee hearing and he was quoted in a magazine in 1950 saying he favoured bombing Russia. Along with Teller and **von Braun**, he was one of the inspirations for the character of Dr Strangelove in the 1964 film.

ERWIN RUDOLF JOSEF ALEXANDER SCHRÖDINGER
12 August 1887 – 4 January 1961

Austrian physicist who was one of the founding fathers of quantum mechanics

'We cannot, however, manage to make do with such old, familiar, and seemingly indispensible terms as "real" or "only possible"; we are never in a position to say what really is or what really happens, but we can only say what will be observed in any concrete individual case. Will we have to be permanently satisfied with this? On principle, yes. On principle, there is nothing new in the postulate that in the end exact science should aim at nothing more than the description of what can really be observed.'

Schrödinger's mulls over the implications of quantum mechanics in his Nobel lecture (12 December 1933)

Schrödinger studied physics at the University of Vienna and remained there conducting practical work for students until the First World War, when he served in the Austrian army as an artillery officer. After several short research posts, he settled at the University of Zurich from 1921 to 1927.

In 1927, Schrödinger moved to Berlin as **Planck**'s successor. He spent time in England and the US after Hitler's rise to power in 1933, but missed his home country and accepted a position at University of Graz in 1936. He feared trouble following the German annexation of Austria in 1938, because his departure from Germany in 1933 was perceived as hostile, so soon afterwards he escaped to Italy. After brief stays in England and Belgium, he became director of the new Institute for Advanced Studies in Dublin, where he remained until his retirement in 1955.

Schrödinger's period at Zurich during the 1920s was the most productive spell of his life. He researched subjects from specific heats of solids to problems in thermodynamics and atomic spectra. But his key discovery came in 1926, when he formulated Schrödinger's wave equation, which neatly predicted the energy levels of electrons in the hydrogen atom. This took account of **de Broglie**'s proposal that particles can behave in a fuzzy wave-like manner, and resolved the problems of **Bohr**'s model of the atom, which represented electrons orbiting an atomic nucleus like point-like planets orbiting the Sun.

For his pioneering work on quantum mechanics, Schrödinger shared the 1933 Nobel prize for physics with **Dirac**. He also attempted to create a unified theory for the gravitational and electromagnetic forces, a feat that defeated him and still remains a problem today.

In popular culture, Schrödinger is best remembered for his famous 'Schrödinger's cat' thought experiment, which he outlined in 1935. He wanted to highlight a problematic paradox in quantum mechanics – that the properties of particles can't be determined until they're observed. The position of an electron, for instance, is a superposition of all possible positions until it is measured.

The thought experiment questions what would happen to a cat that was shut in a box with a device containing a radioactive nucleus and a lethal poison. If the radioactive nucleus decayed by emitting a particle, that would trigger the release of the poison, instantly killing the unfortunate moggie. But quantum mechanics forbids any predictions of when the nucleus will decay. Does this mean the cat is both dead and alive, until somebody 'measures' its state by opening the box and looking at it?

To this day, scientists debate various solutions to this paradox. Perhaps the simplest view is that quantum theory doesn't really create this paradox at all, because it clearly states that the only possible measurements are sensible ones, in this case a cat that is dead or alive, nothing in between.

WERNER KARL HEISENBERG
5 December 1901 – 1 February 1976

German physicist who pioneered quantum theory and formulated the uncertainty principle

'The quantum theory indicates that an understanding of those still unclarified features of atomic physics can only be acquired by foregoing visualization and objectification to an extent greater than that customary hitherto. We have probably no reason to regret this, because the thought of the great epistemological difficulties with which the visual atom concept of earlier physics had to contend gives us the hope that the abstracter atomic physics developing at present will one day fit more harmoniously into the great edifice of science.'

From Heisenberg's Nobel lecture (11 December 1933)

Heisenberg was born in Würzburg and studied physics at the universities of Munich and Göttingen, where he was taught by many eminent scientists including **Born** and **Hilbert**. He earned his PhD at Munich in 1923 before becoming assistant to Born at Göttingen. From 1924 until 1927 he worked with **Bohr** at the University of Copenhagen in Denmark until he was appointed professor of physics at the University of Leipzig aged just 26.

In 1941, Heisenberg became professor at the University of Berlin and director of the Kaiser Wilhelm Institute for Physics. Following the Second World War, he was taken prisoner by the American army and sent to England. On his return in 1946, he helped reorganize the Institute for Physics at Göttingen (later renamed the Max Planck Institute for Physics). He moved with this institute to Munich, where he became professor at the university in 1958.

Heisenberg's work on quantum theory stressed the need to move away from Bohr's description of the atom as electrons orbiting an atomic nucleus like planets orbiting the Sun – it is impossible to assign to an electron an exact position in space at a given time or track its orbit. He also showed that mechanical quantities such as position and velocity were most fruitfully represented in quantum theory by matrices of numbers and expressions.

From this work and studies of radiation emitted by hydrogen, Heisenberg predicted that hydrogen exists in two allotropes – 'ortho-' and 'para-hydrogen' – in which the two nuclei of the atoms in a hydrogen molecule spin in the same or opposite directions respectively. The allotropes were discovered in 1929.

But Heisenberg is probably best known for his uncertainty principle, published in 1927, which underlines the fuzziness of the quantum world. It states that certain pairs of properties, such as the position and momentum of a particle, can't both be determined with exact precision at the same time. The more precisely you know the particle's position, the less precisely you can know its momentum.

The principle stems from the wave-like behaviour of a particle. The only wave with a definite position is one that's concentrated at a single point, but such a wave has an indefinite wavelength, meaning it also has an indefinite momentum. Conversely, the only wave with one precise wavelength is infinitely long, and therefore has no definite position. So in quantum mechanics, there are no states that simultaneously describe a particle's precise position and momentum.

Heisenberg's uncertainty principle quantifies this curious vagueness – the product of the uncertainty in a particle's position and the uncertainty in its momentum must be greater than or equal to 'Planck's constant' (a tiny number equal to 6.6×10^{-34} joule seconds) divided by 4π. For his work on quantum mechanics, Heisenberg won the 1932 Nobel prize for physics. From the 1950s, he branched out into broader research on elementary particles as well as plasma physics and thermonuclear processes.

PAUL ADRIEN MAURICE DIRAC
8 August 1902 – 20 October 1984

*British mathematician and physicist who developed
the theory of quantum mechanics and predicted the
existence of antimatter*

'From general philosophical grounds one would at first sight like
to have as few kinds of elementary particles as possible, say only
one kind, or at most two, and to have all matter built up of these
elementary kinds. It appears from the experimental results,
though, that there must be more than this. In fact the number
of kinds of elementary particle has shown a rather alarming
tendency to increase during recent years.'

From Dirac's Nobel lecture (12 December 1933)

Dirac was born to a Swiss father and
English mother in Bristol, England. He did a degree
in electrical engineering at Bristol University,
graduating in 1921, and then studied maths there for
two years before doing a PhD in maths at Cambridge
University. He received his PhD in 1926. After that
he remained for most of his career at Cambridge,
where in 1932 he was appointed Lucasian Professor
of Mathematics, a post previously held by **Newton** and later by **Hawking**.
In 1969, he moved to the US and later continued his research as professor
of physics at Florida State University.

Dirac began working on the theory of quantum mechanics as soon as it
was introduced in 1925 by **Heisenberg**. While **Schrödinger** had developed
wave equations to describe the behaviour of electrons, these equations
were non-relativistic. Dirac developed wave equations that were consistent
with **Einstein**'s theory of special relativity, and they accurately predicted
the energy levels of a hydrogen atom.

In this work, he also predicted the existence of antimatter. His
mathematical descriptions of the electron suggested there must exist a
second particle in nature, identical to the electron but with opposite electric
charge – the anti-electron or 'positron'. If an electron met a positron, the
two would vanish in a puff of radiation. Likewise, every fundamental

particle in nature should have an antimatter counterpart – antiprotons and antineutrons should exist, capable of building the nuclei of anti-atoms.

Following the first experimental discovery of the positron in 1932, Dirac shared the 1933 Nobel prize for physics with Schrödinger. In his Nobel lecture, Dirac insightfully pointed out an intriguing puzzle that remains unsolved to this day. Antimatter has just as much right to exist in the Universe as normal matter. So why is our solar system, including the Earth and everybody on it, made of matter only?

Dirac suggested that matter and antimatter might even out on large scales – perhaps distant stars are made of antimatter. Antimatter stars would emit the same kind of light as normal ones, so you couldn't tell the difference just by looking at them. The issue remains baffling today. Either there are distant antimatter realms in the Universe, or there's an asymmetry in the laws of nature which meant that matter naturally dominated over antimatter following the Big Bang that created the Universe.

Dirac made many other important steps in quantum mechanics and in 1931 proposed the existence of the magnetic monopole – a particle that carries an isolated north or south magnetic pole. Experimental searches for these elusive particles have so far not found any, however. He is also regarded as the founder of quantum electrodynamics, being the first to use that term; this was later developed by **Feynman** and others. He wrote several books including *Quantum Theory of the Electron* (1928) and *The Principles of Quantum Mechanics* (1930), which has remained one of the standard textbooks on quantum mechanics even now.

A popular biography of Dirac, *The Strangest Man: The Hidden Life of Paul Dirac, Quantum Genius* (2009) by Graham Farmelo, portrays him as an odd character who was aloof and inflexible, possibly autistic, and not a man of many words. One anecdote describes what happened when two graduate students visited him to discuss their ideas about quantum mechanics:

'During their fifteen minute presentation, Dirac said nothing. Afterwards, the students braced themselves for his perceptive comments, but there was an agonisingly long silence, eventually broken by Dirac when he asked them, "Where is the post office?" The students offered to take him there and suggested that he could tell them what he thought of their presentation. Dirac told them, "I can't do two things at once".'

ALEXANDER FLEMING
6 August 1881 – 11 March 1955

Scottish biologist and pharmacologist who recognized the antibiotic effect of penicillin, a drug that has saved countless lives

'I have been trying to point out that in our lives chance may have an astonishing influence and, if I may offer advice to the young laboratory worker, it would be this – never to neglect an extraordinary appearance or happening'

Fleming's advice during a lecture at Harvard University

Born in Ayrshire, Fleming went to school in Kilmarnock and then attended the London Polytechnic Institute, which he left at the age of 16 to become a shipping clerk. In 1901, an uncle left him a legacy that enabled him to study medicine, encouraged by his older brother Tom, a physician. So he enrolled for a medical degree at St Mary's Hospital Medical School in Paddington, recognized as a school of the University of London in 1900, and qualified in 1906.

Fleming then began a research career at St Mary's, interrupted by the First World War, during which he served as a captain in the Royal Army Medical Corps and worked in battlefield hospitals in France. After the war he returned to St Mary's, where he was promoted to professor of bacteriology in 1928, remaining there for the rest of his working life.

During the war, Fleming witnessed the impact of infected wounds on soldiers, which killed many more of them than direct enemy action. He showed that the chemical antiseptics used to treat infected wounds at the time often did more harm than good – they failed to reach bacteria in deep wounds and counterproductively also suppressed a patient's natural immune defences.

By chance in 1921, Fleming made a key discovery when he was suffering from a cold and on a whim cultured some of his nasal mucus on a petri dish. A few weeks later, he found that bacteria were thriving in the dish, except in the blob of mucus. Further experiments showed that something in the mucus seemed to kill some bacteria, but not all of them.

Further work suggested to Fleming that the ingredient must be an enzyme, which he and his colleagues named lysozyme, that he also found in bodily fluids of a wide range of animals including rabbits and dogs. Lysozyme did not kill any of the bacteria that cause serious disease in people. But Fleming had nonetheless demonstrated that it is possible to find a natural substance that selectively kills bacteria without harming normal animal cells.

His finest hour was yet to come in 1928, when he accidentally left a dish of staphylococcus bacteria uncovered for a few days. He returned to find the dish dotted with bacterial growth, apart from one area where a patch of mould from the genus *Penicillium* was growing. The mould produced a substance, later named penicillin by Fleming, which inhibited bacterial growth and turned out to effectively inhibit a wide range of harmful bacteria, including the pathogens that cause pneumonia, scarlet fever and meningitis.

Fleming published a report on penicillin in 1929, but it sparked little interest. Cultivation of the *Penicillium* mould was difficult and he couldn't isolate the antibiotic agent in large quantities. He continued to work on the problem until 1940. Soon after, **Florey** and **Chain** at the Radcliffe Infirmary in Oxford took up the gauntlet and with funding from the British and US governments began mass producing it. During the Second World War, penicillin was available for all wounded Allied forces by D-Day.

By the middle of the 20th century, Fleming's discovery had spurred the large-scale production of synthetic antibiotics that could conquer some of humanity's ancient scourges, including tuberculosis, syphilis and gangrene. For his work on penicillin, Fleming shared the 1945 Nobel prize for physiology or medicine with Florey and Chain.

In a speech he gave for the Nobel banquet on 10 December 1945, Fleming took the opportunity to identify two important factors in such successful research. One was individual freedom. 'Team work may inhibit the primary initiation of something quite new but once a clue has been obtained team work may be absolutely necessary to bring the discovery to full advantage,' he said. Likewise, chance intervened; the suspension of normal financial restrictions and horrendous casualties during the Second World War were a huge driver of the drug's development and large-scale production. He concluded: 'It may be that while we think we are masters of the situation, we are merely pawns being moved about on the board of life by some superior power.' Fleming was knighted in 1944.

INGE LEHMANN
13 May 1888 – 21 February 1993

Danish seismologist who predicted the existence of the Earth's inner core

'I remember Inge one Sunday in her beloved garden ... with a big table filled with cardboard oatmeal boxes. In the boxes were cardboard cards with information on earthquakes ... all over the world. This was before computer processing was available, but the system was the same. With her cardboard cards and her oatmeal boxes, Inge registered the velocity of propagation of the earthquakes to all parts of the globe. By means of this information, she deduced new theories of the inner parts of the Earth.'

Lehmann's nephew Niles Groes recalls his aunt's work

Lehmann studied maths at the universities of Copenhagen and Cambridge. She worked for insurance companies before returning to the University of Copenhagen, where she earned a PhD in 1928. After that, she became head of seismology at Geodetical Institute of Denmark and was responsible for the operations of seismological stations in Greenland.

In 1929, when a large earthquake occurred near New Zealand, Lehmann was puzzled by the reflections of P-waves (primary or compressional waves). She correctly reasoned that some of them travelled some distance into the Earth's core before being reflected by a previously unknown boundary. That built the foundations of a 1936 report in which she correctly proposed that the Earth's centre consists of two parts: a solid inner core surrounded by a liquid outer core.

She also discovered a second discontinuity, now called the Lehmann discontinuity, at a depth of about 220 km (140 miles). The nature of this discontinuity and whether it exists globally is still a matter of debate.

HERMANN JULIUS OBERTH
25 June 1894 – 28 December 1989

German physicist and engineer who opened up the path to the space age

'To make available for life every place where life is possible. To make inhabitable all worlds as yet uninhabitable, and all life purposeful.'

Oberth describes his ultimate ambition
(quoted in a NASA article for educators at www.nasa.gov)

Oberth was born to a Saxon family in Hermannstadt, Transylvannia, then in Austria–Hungary (now Sibiu, Romania). Inspired by Jules Verne's science fiction descriptions of space travel, he drew designs for rockets and built models. He studied medicine in Munich from 1912 then was drafted into the Imperial German Army during the First World War as a doctor. In his spare time, however, he continued his work on rocketry and after the war, he studied physics in Munich and Göttingen.

After Oberth's proposed doctoral thesis on rocketry was rejected, he expanded his ideas in the book *Ways to Spaceflight* (1929). This explained how rockets could escape the Earth's gravitational pull and finally gained Oberth widespread recognition. He foresaw just about all space projects that eventually came to fruition, including Earth observation satellites, interplanetary probes, astronauts landing on the Moon and manned space stations. He also recognized the power of multistage rockets, writing: 'If there is a small rocket on top of a big one, and if the big one is jettisoned and the small one is ignited, then their speeds are added.'

Oberth launched his first liquid-fuel rocket in 1929. Around this time he became a mentor for **von Braun**, who headed the development of German V-2 ballistic missiles during the Second World War. The design utilized many of Oberth's inventions and suggestions. After the war, he worked as an independent rocketry consultant and writer, and from 1955 to 1959 he worked for his former assistant von Braun at NASA's Marshall Space Flight Center in Alabama. While **Tsiolkovsky** and **Goddard** conducted similar research and arrived at similar conclusions, they all did so independently.

ENRICO FERMI
29 September 1901 – 28 November 1954

Italian–American physicist who developed atomic theory and quantum mechanics

'Such a weapon goes far beyond any military objective and enters the range of very great natural catastrophes. By its very nature it cannot be confined to a military objective but becomes a weapon which in practical effect is almost one of genocide. It is clear that the use of such a weapon cannot be justified on any ethical ground which gives a human being a certain individuality and dignity even if he happens to be a resident of an enemy country ... The fact that no limits exist to the destructiveness of this weapon makes its very existence and the knowledge of its construction a danger to humanity as a whole.'

Fermi and his colleague **Rabi** voice opposition to the hydrogen bomb in a report for the General Advisory Committee for the Atomic Energy Commission (1949)

Born in Rome, Fermi studied at the Scuola Normale Superiore in Pisa and at the University of Pisa, where he earned a doctorate in physics in 1922. Funded by scholarships, he then worked with **Born** in Göttingen, Germany, and in the Netherlands before becoming a lecturer at the University of Florence. In 1927, he was appointed as professor of theoretical physics at the University of Rome.

In 1938, Fermi emigrated to the US and became professor at Columbia University in New York City. He later worked on the Manhattan Project to develop nuclear weapons and became a naturalized American citizen in 1944. Following the Second World War, he became professor at the Institute for Nuclear Studies of the University of Chicago, a position that he held until his death.

In 1926, Fermi discovered the statistical laws, now known as Fermi statistics, that govern particles with the quantum property of half-integer spin ('fermions'). Fermions include the electrons, protons and neutrons that make up normal atoms and unlike bosons (see **Bose**), they are subject to **Pauli**'s exclusion principle, which says that no two identical fermions can occupy the same quantum state simultaneously. This underlies the stability and chemical behaviour of atoms, preventing electrons falling into their lowest possible energy states.

In the 1930s, Fermi clarified the nature of radioactive beta decay by focusing on the atomic nucleus as the origin of the emission of an electron and invoking the release of the neutrino, a hypothetical particle proposed by Pauli and presumed to have zero electric charge. Fermi himself coined the term neutrino, Italian for 'little neutral one'. After the discovery of artificial radioactivity by **Jean Frédéric** and **Irène Joliot-Curie**, he also demonstrated that nuclear transformation occurs in almost every element subjected to neutron bombardment.

When **Hahn** and **Meitner** discovered nuclear fission in the late 1930s, Fermi quickly realized it would be possible to sustain fission in a chain reaction. Each fission reaction would release slow neutrons that could trigger another fission reaction. He led vigorous efforts to develop research that ultimately led to the first controlled nuclear chain reaction, which took place on 2 December 1942 at a rackets court underneath a football stadium in Chicago.

Fermi went on to become one of the leading players in the Manhattan Project to develop nuclear fission weapons, which were dropped on Japan in 1945. However, he strongly opposed the subsequent development of the more powerful hydrogen (nuclear fusion) bomb following the detonation of the first Soviet fission bomb in August 1949.

During an informal discussion in 1950, he raised an issue now known as the 'Fermi paradox' – the apparent size and age of the Universe suggest that many technologically advanced extraterrestrial civilizations should exist, but if they do, why don't we know about it? He suggested we should have seen some sort of observational evidence, such as visiting spacecraft. This prompted debate on possible answers, including the 'galactic zoo' hypothesis, that intelligent civilizations in our Milky Way galaxy do exist and observe the Earth, but choose not to make contact for reasons of their own.

For his work on nuclear physics, Fermi won the 1938 Nobel prize for physics. He died in Chicago from stomach cancer, aged 53, in 1954. The element with atomic number 100 was named fermium. The US Fermi National Accelerator Laboratory (Fermilab) near Chicago and the Fermi Gamma-ray Space Telescope, launched in 2008, were also named in his honour.

ISIDOR ISAAC RABI
29 JULY 1898 – 11 JANUARY 1988

Galician-born American physicist who discovered nuclear magnetic resonance, leading to new medical imaging techniques

'It was eerie ... I saw myself in that machine. I never thought my work would come to this.'

Rabi recalls seeing a distorted reflection of his own face inside an MRI scanner a few weeks before his death (from *Rabi: Scientist and Citizen* by John S. Rigden, 1987)

While still a baby, Rabi's family emigrated from Rymanów in Austria-Hungary (now in Poland) to the US. Here he studied chemistry at Cornell University, earning his first degree in 1919. For work on the magnetic properties of crystals, he earned his PhD from Columbia University, New York, in 1927.

Funded by fellowships, he spent two years in Europe working with scientific luminaries such as **Bohr**, **Pauli** and **Heisenberg**. He was appointed professor at Columbia University in 1937, and he remained there until his retirement apart from a spell during the Second World War, when he worked on development of radar and atomic weapons.

Rabi is best known for his discovery of nuclear magnetic resonance (NMR), in which atomic nuclei in a magnetic field absorb and re-emit radio waves. The emitted radiation is at a specific frequency that depends on the magnetic field strength as well as the magnetic properties of the particular isotope emitting it. Different isotopes resonate at different frequencies, the magnetic field strength being equal.

This led to radically new ways of investigating the structure of molecules as well as advanced medical imaging techniques such as MRI (magnetic resonance imaging). For his work, Rabi was awarded the 1944 Nobel prize for physics.

CLYDE WILLIAM TOMBAUGH
4 February 1906 – 17 January 1997

American astronomer who discovered Pluto

'Behold the heavens and the great vastness thereof, for a planet could
be anywhere therein ... Thou shalt dedicate thy whole being to the
search project with infinite patience and perseverance ... Many false
planets shall appear before thee, hundreds of them, and thou shalt
check every one with a third plate ... Thou shalt not engage in any
dissipation, that thy years may be many for thou shalt need them to
finish the job!'

From Tombaugh's 'Ten Commandments for a would-be planet hunter'
(from a 1989 talk in San Francisco)

Tombaugh was born in Streator, Illinois. From the
mid-1920s he built his own telescopes and sent drawings of the planets
he observed to Lowell Observatory in Arizona, where he was appointed
as an assistant in 1929. In 1930, his observations revealed the planet
Pluto, already predicted to exist to explain gravitational disturbances of
Neptune's orbit (although later it became clear that Pluto wasn't really
required to explain this).

Tombaugh went on to study astronomy at the University of Kansas
and during the Second World War he taught navy personnel navigation
at Northern Arizona University. He later worked at White Sands Missile
Range and taught astronomy at New Mexico State University from 1955
until his retirement in 1973.

Pluto was considered the ninth planet in the solar system from its
discovery until 2006, when it was downgraded to 'dwarf planet' status
by the International Astronomical Union. By that time, astronomers had
discovered several bodies the size of Pluto in the outer fringes of the solar
system and many more are likely to turn up in future. Some of Tombaugh's
ashes were placed in the New Horizons spacecraft, launched in 2006 for a
flyby of Pluto in 2015.

LINUS CARL PAULING
28 FEBRUARY 1901 – 19 AUGUST 1994

American chemist who revolutionized chemical bonding theory and campaigned against nuclear weapons

'We are privileged to have the opportunity of contributing to the achievement of the goal of the abolition of war and its replacement by world law. I am confident that we shall succeed in this great task; that the world community will thereby be freed not only from the suffering caused by war but also, through the better use of the Earth's resources, of the discoveries of scientists, and of the efforts of mankind, from hunger, disease, illiteracy, and fear; and that we shall in the course of time be enabled to build a world characterized by economic, political, and social justice for all human beings and a culture worthy of man's intelligence.'

From Pauling's Nobel peace prize lecture (11 December 1963)

Pauling was born in Portland, Oregon, and studied chemical engineering at Oregon Agricultural College in Corvallis (now Oregon State University) where he graduated in 1922. In 1925, he earned a PhD in chemistry and mathematical physics from the California Institute of Technology in Pasadena, where he held posts for most of his career.

Pauling's key contribution was to revolutionize theories about the structures of molecules using **de Broglie**'s wave theory of particles. In 1931, he published a classic paper on the nature of chemical bonding, in which he presented six rules for the shared electron bond. The first three restated earlier observations that the electron-pair bond was formed through the interaction of an unpaired electron on each of two atoms, that the 'spins' (a quantum property) of the electrons had to be opposed, and that once paired, the two electrons could not take part in additional bonds.

The last three rules were new ones. Pauling recognized that the electron-exchange terms for a shared-electron bond involved only one wave function from each atom and that the strongest bonds form between electrons in their lowest energy levels. His final rule about bonding allowed the calculation of bond angles to clarify molecular structures. From the principles of quantum mechanics, he then derived everything from the arrangements of bonds to a complete theory of magnetism in molecules.

During the Second World War, Pauling was a consultant for the explosives division of the US National Defense Research Commission and worked on various projects including developing rocket propellants and oxygen-deficiency alarms for submarines and aircraft. Following the atomic bomb explosions in Japan, Pauling became an active peace campaigner. He calculated and publicized rates of congenital deformity from the fallout of nuclear testing and led a campaign that resulted in a petition signed by more than 9,000 scientists protesting against further testing, which was presented to the UN in 1958. This played an influential role in securing the Nuclear Test Ban Treaty of 1963, which outlawed all nuclear tests except underground ones.

Over his lifetime, Pauling achieved a huge range of scientific breakthroughs, rebuilding chemistry on the new foundation of quantum physics and making key discoveries in X-ray crystallography, immunology, biochemistry and physical chemistry. He also triggered controversy, however, by promoting the fringe view that high doses of vitamin C could successfully treat a range of ills from the common cold to cancer.

For his work on chemical bonding, Pauling won the 1954 Nobel prize for chemistry, and for his anti-nuclear campaigning, he won the Nobel peace prize in 1963. He remains the only person to have won both a peace and science Nobel, and the only person to have won two Nobels that were not shared with anyone else.

JAMES CHADWICK
20 October 1891 – 24 July 1974

English physicist who discovered the neutron and ushered in the modern era of nuclear physics

'I intended to read mathematics at the university. I had no intention whatever of reading physics, but one had to go up for an interview before the term started, and many of them were held in a large hall ... It was divided off into kind of cubicles by low benches ... Mathematics and physics were close together. I sat on the wrong bench.'

Chadwick describes why he studied physics by mistake, an error that he was too shy to correct (from a 1969 interview)

Born in Bollington, Cheshire, Chadwick earned a degree in physics from Manchester University in 1911 then spent the next two years working on radioactivity with **Rutherford**. In 1913 he went to Berlin to work with Hans Geiger (co-inventor of the Geiger counter) but following the outbreak of the First World War, he was interned at a prisoner of war camp near Berlin. After the war, he returned to the UK to work again under Rutherford, who was now at the University of Cambridge.

Rutherford had just succeeded in that year in triggering nuclear fission (splitting the atom) by bombarding nitrogen with alpha particles (helium nuclei). He discovered and named the proton in these experiments. In Cambridge, Chadwick joined Rutherford to continue these experiments with other elements. His most famous result came in 1932, when Chadwick proved the existence of neutrons, nuclear particles similar in mass to protons but devoid of any electrical charge. Rutherford had predicted their existence in 1920, reasoning that if all the mass of the nucleus consisted of protons, the surrounding electrons wouldn't contribute enough negative charge to make the atom neutral overall.

Other physicists including **Irène** and **Frédéric Joliot-Curie** had noticed that when they bombarded light metal beryllium with alpha particles, the beryllium emitted very penetrating radiation that they assumed was gamma rays. Chadwick and Rutherford became convinced they were in fact neutrons. Chadwick repeated these experiments and showed that the best way to explain them was that the alpha particles were knocking neutral nuclear particles out of beryllium and the particles had a mass similar to that of the proton. He reported his discovery in *Nature* on 27 February 1932.

Neutrons can easily penetrate the positively charged nuclei of even very heavy elements, making them disintegrate. So Chadwick's work paved the way for the discovery of uranium-235 fission by **Hahn** and **Meitner**, and the invention of the atomic bomb. It also added clarity to atomic structure, explaining the existence of isotopes, multiple varieties of the same element with different numbers of neutrons in their nuclei.

For his discovery of the neutron, Chadwick won the 1935 Nobel prize for physics. In that year, he was appointed professor at the University of Liverpool, where he built the UK's first cyclotron, a particle accelerator that uses magnetic fields to accelerate charged particles on circular paths. He later became head of the British mission to the Manhattan Project in the US and was knighted in 1945. From 1957 to 1962, he was also a member of the United Kingdom Atomic Energy Authority.

Colleagues have described Chadwick as a shy and retiring man, who could feel physically ill at the thought of a difficult meeting. But he nonetheless proved in his high-level atomic energy work to be an excellent diplomat and people manager.

J(OHN) B(URDON) S(ANDERSON) HALDANE
5 November 1892 – 1 December 1964

British–Indian biologist who pioneered population genetics

'My practice as a scientist is atheistic. That is to say, when I
set up an experiment I assume that no god, angel, or devil is
going to interfere with its course; and this assumption has been
justified by such success as I have achieved in my professional
career. I should therefore be intellectually dishonest if I were
not also atheistic in the affairs of the world.'

From J.B.S. Haldane's *Fact and Faith* (1934)

The son of physiologist John Scott Haldane,
Haldane was born in Oxford. From a young age he
assisted in his father's experiments. He studied
maths, classics and philosophy at Oxford University
then joined the Black Watch regiment during the
First World War. He returned to Oxford after the war
to study physiology, and in 1923 he moved to work
under **Hopkins**. He was then professor at University
College London before in 1957 emigrating to India, where he became a
naturalized citizen in 1961.

Haldane is best known for his quantitative work on genetics. In 1932,
he made the first estimate for the mutation rate of a human gene. He also
introduced the concept of genetic load, the percentage of harmful or lethal
genes in a species population. His books included *The Causes of Evolution*
(1932), which firmly established natural selection as the main premier
mechanism of evolution by explaining it mathematically as a consequence
of Mendelian genetics (see **Mendel**).

He also investigated the physiology of breathing, particularly with
respect to deep-sea diving and safety in mines. Beyond his scientific
research, Haldane was an outspoken public figure whose frequent
pronouncements on subjects ranging from meteorites to economics and
politics were widely publicized.

KARL GUTHE JANSKY
22 October 1905 – 14 February 1950

American radio engineer who first discovered radio emissions from outer space

'Electromagnetic waves of an unknown origin were detected during a series of experiments on atmospherics at high frequencies ... The direction of arrival changes approximately 360 degrees in about 24 hours in a manner that is accounted for by the daily rotation of the Earth ... The waves come from some source outside the solar system.'

From Jansky's classic paper reporting his discovery (*Proceedings of the IEEE*, July 1998)

Born in Norman, Oklahoma, Jansky studied physics at the University of Wisconsin before in 1928 joining the staff of Bell Telephone Laboratories in Holmdel, New Jersey. At Bell Labs, Jansky was tasked with building an antenna to investigate possible sources of static that might interfere with transatlantic radio communications.

Using his antenna, Jansky identified thunderstorms as a source of interference as well as a faint steady hiss of unknown origin. He showed that this hiss rises and falls over 23 hours and 56 minutes, the time it takes for distant stars to return to the same point in the sky due to the Earth's rotation. Eventually, he concluded that the radio emission comes from the Milky Way galaxy itself, the strongest source being towards the galactic centre.

Jansky was keen to build a larger antenna to investigate the Milky Way's radio emissions in more detail, but Bell Labs assigned him to other projects. Radio astronomy didn't take off until American amateur astronomer Grote Reber and others began mapping the sky at radio wavelengths from the late 1930s. Jansky died of kidney disease aged only 44. A unit of flux density widely used by radio astronomers is named after him.

HAROLD CLAYTON UREY
29 April 1893 – 5 January 1981

American chemist who did pioneering work on isotopes and the Earth's early history

'Above all, I regret that scientific experiments – some of them mine – should have produced such a terrible weapon as the hydrogen bomb. Regret, with all my soul, but not guilt.'

Urey quoted in the *New York Times* (27 April 1961)

Urey studied zoology at the University of Montana, earning a bachelor's degree in 1917 and then spent two years as a research chemist. In 1923, he earned a PhD in chemistry from the University of California, Berkeley. After holding a post at Johns Hopkins University in Maryland, he held professorships at Columbia University, the University of Chicago and the University of California, San Diego.

In 1931, Urey developed a method for concentrating any possible heavy hydrogen isotopes (hydrogen with one or more neutrons in the nucleus) in liquid hydrogen. This led him to discover the isotope deuterium. For this work, he won the 1934 Nobel prize for chemistry. He later worked on the separation of uranium isotopes, which played a key role in the Manhattan Project to develop nuclear weapons.

Urey also had diverse research interests including the origins of the planets and the conditions on the early Earth. He speculated that the early terrestrial atmosphere was probably composed of ammonia, methane and hydrogen. In a famous test called the Miller–Urey experiment, Urey's student **Miller** showed that if such a mixture is exposed to water and sparks, the analogue of lightning on the young Earth, it can react to produce amino acids, the building blocks of life.

PAVEL ALEKSEYEVICH CHERENKOV
28 JULY 1904 – 6 JANUARY 1990

*Soviet physicist who investigated Cherenkov radiation,
an eerie blue glow emitted by particles moving faster
than light in a medium such as air or water*

'We were dealing with no ordinary everyday luminescence,
but with a phenomenon of an entirely new kind, and one of
extraordinary interest not only on account of its significance
in principle but also in regard to the many practical
possibilities for its use'

From Cherenkov's Nobel lecture (11 December 1958)

Cherenkov was born to a peasant family
in the Voronezh region of Russia. He graduated in
physics and maths at Voronezh State University in
1928, and in 1930 became a senior scientific officer
at the P.N. Lebedev Institute of Physics in Moscow,
where he was appointed professor in 1953.

In 1934, working under Sergey Ivanovich
Vavilov, Cherenkov observed a strange blue glow
coming from a bottle of water bombarded by radioactive particles. His
colleagues Ilya Mikhailovich Frank and Igor Yevgenyevich Tamm correctly
concluded that the glow is due to particles moving through the water faster
than the speed of light in water, the visible equivalent of a sonic boom.
(This doesn't contradict **Einstein**'s conclusion that no object can move
faster than the speed of light in a vacuum; in water, light's speed is only
about three quarters of the vacuum value.)

Further experiments by Cherenkov verified this theory's predictions.
But when he sent a report on it to the British journal *Nature* for publication,
Nature rejected it. That turned out to be a poor call – Cherenkov, Frank
and Tamm went on to share the 1958 Nobel prize for physics for their
work. Cherenkov radiation has since played a key role in the detection of
high-energy particles from space.

IRÈNE JOLIOT-CURIE
12 September 1897 – 17 March 1956

(JEAN) FRÉDÉRIC JOLIOT-CURIE
19 March 1900 – 14 August 1958

French physicists who created the first artificial radioactive elements

'It was indeed a great source of satisfaction for our lamented teacher Marie Curie to have witnessed this lengthening of the list of radio-elements which she had had the glory, in company with Pierre Curie, of beginning. The diversity of the chemical properties, diversity of the average lives of these synthetic radio-elements will without doubt enable further advances in research in biology and in physical chemistry to be made.'

From Frédéric Joliot's Nobel lecture (12 December 1935)

Irène Joliot-Curie (pictured), the daughter of **Marie** and **Pierre Curie**, was born in Paris. She benefited from wide-ranging private tutoring after her mother teamed up with eminent French academics including **Langevin** to educate all their children in their homes. After studying at the University of Paris, Irène served as a radiographer in field hospitals during the First World War.

Afterwards, Irène Curie returned to Paris to study at the Radium Institute established by her parents and she earned a doctorate in 1925 for a thesis on alpha radiation from the radioactive element polonium, which her parents had discovered. Appointed as a lecturer at the University of Paris in 1932, she became professor there in 1937, and later director of the Radium Institute from 1946.

In 1926, she married Frédéric Joliot, also born in Paris. He studied engineering in Paris and became an assistant to Marie Curie at the Radium Institute in 1925. He earned his doctorate in 1930 for a thesis on the electrochemistry of radioactive elements and was appointed as a lecturer at the University of Paris in 1935. In 1937, he became professor at the Collège de

France. After his wife's death, he also succeeded her as professor of nuclear physics at the University of Paris.

In the 1930s, the Joliot-Curies collaborated on research into the structure of the atom. Their greatest discovery was artificial radioactivity, in 1934. By bombarding boron, aluminium and magnesium with alpha particles, they produced radioactive isotopes of nitrogen, phosphorus, silicon and aluminium. These isotopes are not found in nature and spontaneously decay by emitting electrons or their antimatter counterparts, positrons.

For the first time, it became clear that radioactivity was not just a property of very heavy elements like uranium; any element could be radioactive if the right isotope could be created. Although experiments by the Joliot-Curies indicated the existence of both the positron and the neutron, they did not recognize this and the discoveries of the two particles were later credited to the American physicist Carl Anderson and **Chadwick**.

Their techniques led to the creation of many short-lived radioisotopes that are useful in medicine and can be manufactured quickly, cheaply and plentifully. The Joliot-Curies also researched the theory of nuclear chain reactions and as well as concepts for an atomic pile that would sustain them. On the advance of the German forces in 1940, Frédéric Joliot-Curie arranged for the documents and materials relating to this work to be transported to England. After the war, he and his wife were both appointed as commissioners for the French Atomic Energy Commission and worked on the construction of the first French nuclear reactor in 1948.

For their work synthesizing new radioactive elements, the Joliot-Curies shared the 1935 Nobel prize for chemistry. The Curies remain the most successful 'Nobel prize family', with four members of two generations winning five prizes altogether. Both Irène and Frédéric Joliot-Curie were committed socialists. During the German occupation of France in the Second World War, Frédéric Joliot-Curie took an active part in the Resistance, helped **Langevin** escape to Switzerland and joined the Communist party. The Soviet Union awarded him the International Stalin Prize for Strengthening Peace Among Peoples in 1951.

The Joliot-Curies both died relatively young, Irène from leukaemia aged 58 and Frédéric, also aged 58, from a liver condition. High levels of radiation exposure from their lab experiments probably contributed to their illnesses.

KONRAD ZACHARIAS LORENZ
7 November 1903 – 27 February 1989

Austrian zoologist who founded modern ethology

'Natural causal associations have always turned out to be grander and more awe-inspiring than even the most imaginative mythical interpretation. The true scientist does not need the inexplorable, the supernatural, to evoke his reverence: for him there is only one miracle, namely, that everything, even the finest flowerings of life, have come into being without miracles.'

From Lorenz's *On Aggression* (1963)

Born in Vienna, Lorenz studied medicine at the university there from 1923 and earned his MD in 1928. He went on to study zoology in Vienna, earning a doctorate in 1933. In 1939, he was appointed professor of psychology at the University of Königsberg (now Kaliningrad, Russia). Serving as a doctor in the German army from 1941, he was captured and held as a prisoner of war in the Soviet Union from 1944 to 1948. After returning to Vienna, he cofounded the Institute for Behavioural Physiology near Munich in 1955.

In studies of the behaviour of greylag geese, Lorenz demonstrated the concept of 'imprinting', in which soon after birth, an animal fixates on parental behaviour, for instance by accepting a foster mother in the place of its real mother, even if the foster mother is a different species. He also discovered that animals have instinctive behaviour patterns, such as fighting and surrender postures, that remain dormant until triggered for the first time by an event, and these are not learned but are genetically programmed.

Late in life, he analysed human social behaviour as well. For his work, he won a share of the 1973 Nobel prize for physiology or medicine. He wrote many books, including the popular titles *King Solomon's Ring* (1949) and *On Aggression* (1963).

TROFIM DENISOVICH LYSENKO
29 September 1898 – 20 November 1976

Soviet agronomist who hampered scientific agricultural progress

'Genetics was to be totally suppressed; Lysenko was to be the undisputed boss of all biology and all its agricultural application. That was the ultimate victory of the practical Stalinist approach to scientific agriculture, the final arrival at those windy heights from which the nature of things can be revolutionized by a flick of wilful intuition. It was also ultimate self-defeat.'

From *The Lysenko Affair* by David Joravsky (1986)

Born in Karlovka, Ukraine, Lysenko graduated from the Poltava School of Horticulture in 1921 and from the Kiev Institute of Agriculture in 1925. From 1940 to 1965, he was director of the Soviet Academy of Science's Institute of Genetics.

In the late 1920s, Lysenko showed in experiments at an agriculture station in Azerbaijan that he could cool wheat seeds normally sown in winter and treat them with moisture to make them bear crops when planted in spring. This 'vernalization' would prevent crops being destroyed during harsh winters. He also claimed, falsely, that the trait is inherited by following generations of plants, even though they were not treated. Various other advances in crop breeding he claimed were often impossible to replicate.

Nonetheless, he won support from Soviet political bosses. Denouncing **Mendel**'s theory of genetics as false, Lysenko used his powerful position to ensure that scientists who opposed him were fired or silenced. He was largely responsible for the imprisonment of his predecessor at the Institute of Genetics, the Soviet biologist Nikolai Vavilov, who died of starvation in 1943. Following Stalin's death, Lysenko's influence faded and today much of his experimental work is viewed as fraudulent.

ROBERT ALEXANDER WATSON-WATT
13 April 1892 – 5 December 1973

Scottish engineer who developed radar

'Pity Sir Watson-Watt, strange target of this radar plot
And thus, with others I can mention, the victim of his own invention.
His magical all-seeing eye enabled cloud-bound planes to fly
but now by some ironic twist it spots the speeding motorist
and bites, no doubt with legal wit, the hand that once created it.'

Watson-Watt wrote this poem after a Canadian policeman with a radar gun
gave him a speeding ticket

Watson-Watt was a descendant of James Watt, inventor of the first practical steam engine. He studied engineering at University College, Dundee, graduating in 1912. After research work on radio waves, he joined the Royal Aircraft Factory at Farnborough in 1915, tasked with using radio detection systems to locate thunderstorms and provide warnings to pilots.

From 1924, Watson-Watt worked at radio research stations in Slough and Teddington. With the Second World War looming, the Air Ministry was anxious to find ways to fend off air attacks, and one idea mooted was to use high-frequency radio waves to heat attacking aircraft and detonate the bombs inside them. Watson-Watt's assistant, Arnold ('Skip') Wilkins, showed this was impractical, but he pointed out to Watson-Watt that engineers from the General Post Office had noticed radio signals fluttered when aircraft passed.

With Wilkins, Watson-Watt developed the idea into a detailed proposal by 1935. Soon after, a trial using a BBC radio transmitter successfully detected a Heyford biplane bomber. Watson-Watt then led a team near Felixstowe that designed and installed radar stations along the south and east coasts of England before the war began. It played a key role in preventing Germany achieving air superiority over England during the Battle of Britain. Watson-Watt was knighted in 1942.

ALAN MATHISON TURING
23 June 1912 – 7 June 1954

British mathematician who formalized the theoretical concept of computers

'I believe that at the end of the century the use of words and general educated opinion will have altered so much that one will be able to speak of machines thinking without expecting to be contradicted'

From Turing's paper 'Computing Machinery and Intelligence' (1950)

Born in London, Turing studied maths at Cambridge University from 1931. He graduated in 1934 then earned a PhD from Princeton University in 1938. After a productive spell working on enemy code breaking during the Second World War, he worked at the National Physical Laboratory and the University of Manchester. He also worked at GCHQ, the post-war British intelligence communications centre.

In 1936, Turing outlined his concept of an abstract machine, now called a 'Turing machine', that would shift from one state to another using a set of rules and the input of symbols read from and written to tape. It wasn't intended as practical computing technology, but as a conceptual device to clarify the potential and limitations of mechanical computation. But it did anticipate modern computers by invoking the concepts of automating the application of algorithms, reading data and writing data to memory.

After the outbreak of war, Turing joined the effort to crack enemy codes at Bletchley Park in Buckinghamshire. He played a key role in developing the 'bombe', a machine used to decode messages sent by German Enigma machines, which gave Allied forces an upper hand in defeating Germany.

His 'Turing test', described in 1950, outlines a way to test whether a computer has achieved the equivalent of human intelligence. He was arrested in 1952 for practising homosexuality and ordered to undergo oestrogen injections as a supposed 'treatment'. He returned to academia to study a wide range of scientific problems, but his security clearance was withdrawn, meaning he could no longer work for GCHQ. He died aged 41 from cyanide poisoning, which was probably a deliberate suicide.

LUIS WALTER ALVAREZ
13 June 1911 – 1 September 1988

*American physicist and inventor who pioneered
experimental particle physics*

'Most of us who become experimental physicists do so for two
reasons; we love the tools of physics because to us they have
intrinsic beauty, and we dream of finding new secrets
of nature as important and as exciting as those uncovered
by our scientific heroes'

From Alvarez's Nobel lecture (11 December 1968)

Born in San Francisco, Alvarez studied physics at the University
of Chicago, where he earned his first degree in 1932 and a PhD in 1936.
After that, he spent most of his career at the University of California,
Berkeley. His research was amazingly wide-ranging and he was granted
more than 40 patents for his inventions.

In 1937, Alvarez demonstrated in experiments that atoms can decay
when the nucleus captures an inner orbital electron, converting a nuclear
proton to a neutron and decreasing the element's atomic number by one.
With **Bloch,** he also made the first measurement of the magnetic moment of
the neutron. During wartime research, Alvarez developed radar equipment
including radar-guided aircraft landing technologies and detonators for
setting off the plutonium bomb, the type dropped on Nagasaki, Japan. He
also flew in aircraft as an observer at both the Alamogordo atomic bomb
test of 16 July 1945 and the Hiroshima explosion.

After the war, Alvarez worked in particle physics. He designed and
built a linear particle accelerator at Berkeley and developed large liquid
hydrogen bubble chambers to record particle tracks. For his work on
experimental particle physics, he won the 1968 Nobel prize for physics.
With his son, **Walter Alvarez,** he compiled evidence that the extinction
of the dinosaurs on Earth around 65 million years ago was due to a
devastating comet or asteroid impact.

THEODOSIUS GRYGOROVYCH DOBZHANSKY
24 January 1900 – 18 December 1975

Ukrainian–American biologist who synthesized evolutionary biology with genetics

'Man has always been fascinated by the great diversity of organisms which live in the world around him ... All this diversity is at first sight staggering and bewildering. The greatest achievement of biological science to date is the demonstration that the diversity is not fortuitous. It has not arisen from a whim or caprice of some deity. It is a product of evolution.'

From Dobzhansky's *Genetics and the Origin of Species* (1951 edition)

Dobzhansky was born in Nemyriv, Ukraine (then part of Imperial Russia); and studied at the University of Kiev between 1917 and 1924. After that he moved to Leningrad in Russia to work at a lab studying the genetics of *Drosophila melanogaster* (fruit flies). He emigrated to the US in 1927 having secured a Rockefeller Foundation scholarship and worked with **Morgan** at Columbia University in New York. He continued his work with Morgan at the California Institute of Technology from 1930 to 1940, becoming a naturalized US citizen in 1937. After that he held research and teaching posts at several universities including Columbia and the University of California, Davis.

Dobzhansky showed that genetic variability within a fruit fly population is large, but populations from distant regions are more genetically dissimilar than close regional varieties. He also showed that an animal genome includes many potentially harmful recessive genes that can nonetheless help them adapt to environmental change. His classic work *Genetics and the Origin of Species* (1937) was influential in synthesizing evolutionary biology with genetics and helped establish the notion that gene mutation is a key process in Darwinian natural selection.

WERNHER MAGNUS MAXIMILIAN VON BRAUN
23 MARCH 1912 – 16 JUNE 1977

German rocket developer and champion of space exploration

'While for many years, and on two continents, the more immediate task (and the one for which alone support was available) was to build rockets as weapons of war, our long-range objective has remained unchanged to this very day – the continuous evolution of space flight'

Recollections of von Braun, 1963 (documented by NASA's Marshall Space Flight Center History Office)

Born in Wirsitz, Prussia (now Wyrzysk in Poland), von Braun developed a keen interest in space flight during his teenage years and became involved in a German rocket society around 1929. In 1932, he joined the German army to develop ballistic missiles. During this work he earned a PhD in physics from the University of Berlin in 1934 for his thesis on liquid-fuel rockets.

During the Second World War, von Braun became leader of a project to develop V-2 ballistic missiles at a secret laboratory at Peenemünde on Germany's Baltic coast. A liquid-fuel missile about 14 m (46 ft) long, the V-2 could fly faster than 5,630 km/hr (3,500 mph) to bomb targets 800 km (500 miles) away. It was deployed against European targets from September 1944.

Realizing that Germany couldn't win the war, von Braun engineered the surrender of 500 of his top scientists, along with plans and test vehicles, to the Americans. For 15 years after the war, he developed ballistic missiles for the US army. In 1960, his rocket development centre was transferred to the newly established space agency NASA. Von Braun became director of NASA's Marshall Space Flight Center and chief architect of the Saturn V launcher that would famously propel Americans to the Moon in 1969.

ERNST BORIS CHAIN
19 June 1906 – 12 August 1979

German biochemist who developed the drug penicillin

'As a member of one of the most cruelly persecuted races in the world, I am profoundly grateful to Providence that it has fallen to me, together with my friend Sir Howard Florey, to originate this work on penicillin, which has helped to alleviate the suffering of the wounded soldiers of Britain, the country that has adopted me, and the wounded soldiers of our Allies, among them many thousands belonging to my own race, in their bitter struggle against one of the ... most inhuman tyrannies the world has ever seen'

From Chain's Nobel banquet speech (10 December 1945)

Born to a Jewish family in Berlin, Chain graduated in chemistry from the Friedrich-Wilhelm University in Berlin in 1930 and then researched enzymes at the Charité Hospital in Berlin for three years. In 1933, he emigrated from Nazi Germany to England, where he worked with **Hopkins** at the University of Cambridge. He later held lecturing posts and professorships at Oxford University, the Superior Institute of Health in Rome and Imperial College, London.

His chemistry research covered a wide range of topics including snake venoms and tumour metabolism. But he is best known for work he began in 1938 with **Florey**, a systematic study of antibacterial substances produced by microbes, including penicillin (the first effective antibiotic drug), which **Fleming** had investigated nine years earlier. Chain showed that penicillin contains an enzyme that kills bacteria by breaking down their cell walls and his team carried out early trials of penicillin as a drug. He also proposed the structure of penicillin, which was later confirmed by **Hodgkin**.

For his work on penicillin, Chain shared the 1945 Nobel prize for physiology or medicine with Fleming and Florey. He was knighted in 1969.

HOWARD WALTER FLOREY
24 September 1898 – 21 February 1968

Australian pathologist and pharmacologist who developed the drug penicillin

'My colleagues and I have been very fortunate in that we have worked during the last few years on something which has proved to be of some immediate value to mankind ... During the last few years the demonstration of what the application of scientific methods can achieve has been so striking and of such a magnitude that even those brought up in the classical tradition, who form most of the statesmen and politicians of the world, are at last aware of the tremendous tasks that lie ahead in the utilization of these forces.'

From Florey's Nobel banquet speech (10 December 1945)

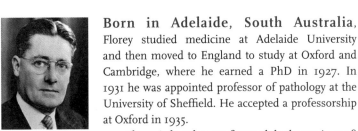

Born in Adelaide, South Australia, Florey studied medicine at Adelaide University and then moved to England to study at Oxford and Cambridge, where he earned a PhD in 1927. In 1931 he was appointed professor of pathology at the University of Sheffield. He accepted a professorship at Oxford in 1935.

Florey is best known for work he began in 1938 with **Chain**, a systematic study of antibacterial substances produced by microbes, including penicillin (the first effective antibiotic drug), which **Fleming** had investigated a decade earlier. Fleming had not managed to isolate the active antibiotic in the moulds he studied. From 1939, Florey and Chain headed a team of British scientists financed by the Rockefeller Foundation who isolated and purified penicillin, and carried out early clinical trials.

This led to successful small-scale manufacture of the drug, and by 1945, enough penicillin was available to treat all the wounded Allied forces on D-Day. Florey shared the 1945 Nobel prize for physiology or medicine with Fleming and Chain, and he was knighted in 1944.

OTTO HAHN
8 March 1879 – 28 July 1968

German physicist who co-discovered nuclear fission

'The discovery of the fission of heavy nuclei has led to consequences
of such a nature that all of us, indeed the whole of humanity, look
forward with great expectations, but at the same time with great
dread, to further developments ... With regard to its practical
application I am also sure that you share all our hopes that this
application will serve in the end as a blessing to mankind.'

Swedish biochemist Arne Tiselius addresses Hahn at a Nobel prize ceremony
(13 December 1946)

Born in Frankfurt, Hahn studied chemistry at Munich and
Marburg, where he earned his PhD in 1901. He became an assistant at
the Chemical Institute in Marburg for two years before briefly working at
University College London under **Ramsay** and then at McGill University
in Montreal, Canada, with **Rutherford**. In 1906 he returned to Germany
to work at **Fischer**'s Chemical Institute at Berlin University. In 1913, he
joined the new Kaiser Wilhelm Institute for Chemistry, where he became
director in 1928.

While working with Ramsay in 1904 to separate radium from an
impure sample of barium chloride, Hahn showed that another radioactive
substance was present and named it radiothorium. In further studies with
Rutherford, he identified two new radioactive species, which he called
thorium-C and radioactinium. In 1907, Hahn began a 30-year collaboration with **Meitner** when she came to Berlin from Vienna.

Hahn and Meitner are best known for experiments that demonstrated
nuclear fission for the first time. Repeating an experiment by **Fermi**, they
bombarded uranium atoms with neutrons and showed that this created
barium, indicating that the experiment had split the atom. For this work,
Hahn won the 1944 Nobel prize for chemistry.

Austrian physicist who co-discovered nuclear fission

'I will have nothing to do with a bomb!'

Meitner's response to being invited to work on the Manhattan Project to develop nuclear
fission weapons (*Lise Meitner: A Life in Physics* by Ruth Lewin Sime, 1997)

Born to a Jewish family in Vienna,
Meitner studied physics at the University of Vienna,
where she earned a doctorate in 1905. She then moved
to Berlin where she attended lectures by **Planck**, who
later appointed her as his assistant. In 1912, she and
Hahn moved to the new Kaiser Wilhelm Institute in
Berlin-Dahlem, and in 1926 she was appointed full
professor at the University of Berlin. In 1938, she
fled to Sweden to escape Nazi persecution, and she remained there until
her retirement in 1960.

Working with Hahn, Meitner discovered several new radioactive
isotopes. Soon after she fled to Sweden, Hahn wrote to her to say he
had discovered radioactive barium as a product of bombarding uranium
with neutrons. Meitner correctly proposed that this was the result of the
uranium nucleus becoming unstable and splitting in two. Although the
possibility of nuclear fission has been suggested before, Meitner and her
nephew Otto Frisch were the first to understand the details of how uranium
would split into barium and krypton, with the emission of neutrons and a
large amount of energy to account for the lower mass of the products.

Meitner was one of few scientists who refused to work on military
projects during the Second World War, turning down an invitation
to contribute to the Manhattan Project to develop the atom bomb. The
element meitnerium (atomic number 109) is named in her honour.

B(URRHUS) F(REDERIC) SKINNER
20 March 1904 – 18 August 1990

American behaviourist who pioneered experimental psychology

'Two exhausting world wars in a single half century have given
no assurance of a lasting peace ... Science has made war more
terrible and more destructive ... It may not be science which is
wrong but only its applications. The methods of science have
been enormously successful wherever they have been tried ...
If we can observe human behavior carefully from an objective
point of view and come to understand it for what it is, we may
be able to adopt a more sensible course of action.'

From Skinner's *Science and Human Behavior* (1953)

Born in Susquehanna, Pennsylvania, Skinner's first
studies were in English, at Hamilton College in upstate New York, but
after an unsuccessful attempt to carve out a writing career, he enrolled
to study psychology at Harvard University. Here, he earned a masters
degree in 1930 and a doctorate in 1931. He remained at Harvard in a
research post until 1936, when he moved to Minneapolis to teach at the
University of Minnesota. From 1948 until his retirement in 1974, he held
professorships at Harvard.

In rat experiments, Skinner showed that behaviour is influenced not
only by a stimulus (such as the association between a bell ringing and food
being supplied, see **Pavlov**) but also by what follows. For instance, if rats
failed to nudge a lever to turn on a light and an electric shock followed,
rats would learn to turn on the light. In this way he established the notion
of reinforcement – behaviour that is reinforced tends to be repeated
(strengthened), while behaviour that is not reinforced tends to die out or
be weakened.

Skinner advocated rigorous scientific study of psychology and carefully
controlled experiments. A prolific writer, he published more than 20 books.

HANS ALBRECHT BETHE
2 July 1906 – 6 March 2005

*German–American physicist who explained how
stars generate their energy*

'From time immemorial people must have been curious to
know what keeps the sun shining ... Stars have a life cycle
much like animals. They get born, they grow, they go through
a definite internal development, and finally they die, to give
back the material of which they are made so that new stars
may live.'

From Bethe's Nobel lecture (11 December 1967)

Born in Strassburg, Germany (now Strasbourg in France),
Bethe studied at the universities of Frankfurt and Munich, where he earned
a PhD in theoretical physics in 1928. He taught in Germany until 1933,
when he lost his job (his mother was Jewish) and emigrated to England,
where he held posts at the universities of Manchester and Bristol. From
1935, he spent most of the rest of his career at Cornell University in Ithaca,
New York, becoming a naturalized citizen of the US in 1941.

During the Second World War, he was head of the theoretical division
at the secret Los Alamos laboratory that developed the first atomic bombs.
But his best-known work was the discovery of the nuclear reactions that
fuse hydrogen into helium in the cores of stars, supplying their energy.
In a 1939 paper *Energy Production in Stars,* he described the proton–proton
chain, a fusion reaction creating intermediary deuterium that drives
energy production in stars with masses up to about that of the Sun.

He also described the carbon–nitrogen–oxygen cycle, in which
stars convert hydrogen to helium using carbon, nitrogen and oxygen as
catalysts. This process dominates in very massive stars. For his work on
nuclear reactions, Bethe won the 1967 Nobel prize for physics.

FELIX BLOCH
23 October 1905 – 10 September 1983

Swiss–American physicist who developed nuclear magnetic resonance techniques

'It is a tribute to the inherent harmony and the organic growth of our branch of science that every advance in physics is largely due to the developments that preceded it'

From Bloch's Nobel lecture (11 December 1952)

Born into a Jewish family in Zurich, Bloch studied engineering at Zurich's Federal Institute of Technology for a year before switching to physics, where he was taught by **Schrödinger**. He then moved to the University of Leipzig, where he studied with **Heisenberg** and earned his doctorate in 1928. Over the following few years, various assistantships and fellowships gave him the opportunity to work with some of the leading lights of physics, including **Pauli**, **Bohr** and **Fermi**.

Bloch left Nazi Germany in 1933 and worked at Stanford University in California from 1934. During the war years, he worked on atomic energy at Stanford and Los Alamos, and later in counter-measures against radar at Harvard University. In 1939, he became a naturalized US citizen.

In the same year, Bloch conducted experiments with **Luis Alvarez** to measure the magnetic moment of the neutron, a property that arises because the neutron has an intrinsic angular momentum, or spin. They did this to an accuracy of 1 per cent. In the 1940s, independently of **Purcell**, he also developed nuclear magnetic resonance techniques for liquids and solids, a method of studying nuclear moments.

The technique involves aligning nuclear moments in a magnetic field, 'wobbling' them with radio waves and then detecting radio signals the nuclei emit as they relax. This is the principle behind magnetic resonance imaging techniques used in modern medicine (see **Mansfield**). For their work on nuclear magnetism, Bloch and Purcell shared the 1952 Nobel prize for physics.

SUBRAMANYAN CHANDRASEKHAR
19 October 1910 – 21 August 1995

Indian astrophysicist who made dramatic predictions about how stars evolve

'The mathematical theory of black holes is a subject of immense complexity, but its study has convinced me of the basic truth of the ancient mottoes: "the simple is the seal of the true" and "beauty is the splendour of truth".'

From Chandrasekhar's Nobel lecture (8 December 1983)

Until the age of 12, Chandrasekhar (his given name) was home schooled and privately educated in Lahore (then in British India). He studied physics at the Presidency College in Madras from 1925 and was later awarded an Indian government scholarship to do a PhD at Cambridge University. He joined the faculty of the University of Chicago in 1937 and remained there for the rest of his life.

Chandrasekhar was one of the earliest scientists to apply the rigours of physics to analyse the evolution of stars. He demonstrated that there is an upper limit – now called the Chandrasekhar limit – to the mass of a white dwarf star, a late stage in the evolution of stars like the Sun. After they run out of fuel, Sun-like stars shrink into hot white balls roughly the size of the Earth. Chandrasekhar showed that if a star in this phase is any heavier than about 1.4 times the mass of the Sun, it will become unstable and either explode or collapse into a much more dense object.

Among other wide-ranging work, Chandrasekhar developed the mathematical theory of black holes, which can form when the core of a very massive star collapses to a point of such high density that nothing, not even light, can escape the intense gravity nearby. He shared the 1983 Nobel prize for physics with **Fowler** and NASA's Chandra X-ray Observatory is named in his honour.

GEORGE WELLS BEADLE
22 October 1903 – 9 June 1989

American geneticist who discovered the role of genes in enzyme production

'First in *Drosophila* [fruit flies] and then in [the bread mould] *Neurospora*, we had rediscovered what Garrod had seen so clearly so many years before ... What Garrod had shown for a few genes and a few chemical reactions in man was true for many genes and many reactions in *Neurospora*.'

From Beadle's Nobel lecture (11 December 1958)

Born in Wahoo, Nebraska, Beadle studied at the College of Agriculture in Lincoln, Nebraska, before graduating from the University of Nebraska in 1926. He earned a PhD in 1931 for work at Cornell University on the genetics of corn plants. From 1931 until 1936 he worked at the California Institute of Technology at Pasadena before moving to Harvard and then Stanford University.

Back in 1909, the English physician Archibald Garrod had proposed that a gene is responsible for the production of a specific enzyme (the chemical nature of DNA and genes was still unknown, however). He suggested correctly that alkaptonuria – an inherited condition that can make urine turn dark brown – results from a single faulty recessive gene, leading to a deficiency in an enzyme that breaks down the chemical homogentisic acid (also called alkapton).

Beadle and his Stanford colleague Edward Lawrie Tatum looked for more general proof of this 'one gene – one enzyme' idea. If it was correct, it should be possible to create genetic mutants that lack specific enzyme reactions. To test this, they irradiated spores of the bread mould *Neurospora crassa* to induce single-gene mutations and found that the mutants had specific nutritional needs. For instance, one couldn't live without extra vitamin B_6.

In this and similar work Beadle and Tatum demonstrated the 'one gene – one enzyme' link and the pair shared half of the 1958 Nobel prize for physiology or medicine.

JOHN ARCHIBALD WHEELER
9 July 1911 – 13 April 2008

American theoretical physicist who developed Einstein's theory of general relativity

'[The black hole] teaches us that space can be crumpled like a piece of paper into an infinitesimal dot, that time can be extinguished like a blown-out flame, and that the laws of physics that we regard as "sacred", as immutable, are anything but'

From *Geons, Black Holes & Quantum Foam: A Life in Physics* (1999)
by Wheeler and Kenneth Ford

Born in Jacksonville, Florida, Wheeler studied physics at Johns Hopkins University in Maryland, where he earned his PhD in 1933. He held a teaching post at the University of North Carolina in Chapel Hill from 1935 until 1938, when he moved to Princeton University in New Jersey for most of the rest of his career. During the Second World War, Wheeler worked on the Manhattan Project to develop atomic weapons and from 1976 until his retirement in 1986, he was director of the Center for Theoretical Physics at the University of Texas.

In 1939, Wheeler worked with **Bohr** on the liquid drop model of nuclear fission. But he is best known for work on mathematical extensions to **Einstein**'s general relativity theory. Wheeler introduced the term 'black hole' to describe a collapsed object so dense that nothing, not even light, can escape the intense gravitational field surrounding it. He also proposed that the whole Universe might undergo a 'big crunch', collapsing into a superdense state like the reverse of the Big Bang and then creating a totally new Universe.

Wheeler introduced the concept of the 'wormhole' to describe hypothetical tunnels in space–time that could act as shortcuts between far-flung regions of the Universe. He also pioneered theories of quantum gravity to try to reconcile quantum theory with general relativity.

EDWARD TELLER
15 January 1908 – 9 September 2003

Hungarian–American theoretical physicist regarded as the father of the hydrogen bomb

'If not for me, the H-bomb would have been developed in Russia first. In the US, we would now be speaking Russian. That I spent my life working on weapons, I have not the least regret. I succeeded. I believe that by building the H-bomb, I contributed to winning the cold war without bloodshed. I am not overly modest.'

From an interview with *Esquire* (31 December 2001)

Born to a Jewish family in Budapest, Teller studied chemical engineering at the University of Karlsruhe and went on to study theoretical physics in Munich and Göttingen. He earned his PhD from the University of Leipzig in 1930 after studying with **Heisenberg**. In 1935, he emigrated to the US and in 1941 he joined the Manhattan Project to develop nuclear fission bombs. He was a cofounder and director of the Lawrence Radiation Laboratory in Livermore (now Lawrence Livermore National Laboratory), established in 1952 for thermonuclear research.

During the development of the atom bomb, scientists feared that an uncontrolled nuclear reaction might continue indefinitely and affect the entire Earth. Calculations by Teller reassured them that the nuclear explosion would only destroy a limited area. In the 1950s, Teller became a vigorous supporter of efforts to develop the hydrogen fusion bomb and he played a leading role in its design. However, he fell out with many of his colleagues after controversially testifying against **Oppenheimer** in a security clearance hearing.

Teller was renowned for his scientific talent but also for his volatile personality and difficult interpersonal relations, and he is considered one of the inspirations for the character of Dr Strangelove in the 1964 film.

SELMAN ABRAHAM WAKSMAN
22 July 1888 – 16 August 1973

Ukrainian-American biochemist who discovered streptomycin, the first antibiotic that could cure tuberculosis

'In learning to utilize antibiotics for the control of human and animal diseases, the medical and veterinary professions have acquired powerful tools for combating infections and epidemics ... Even persons afflicted with those forms of tuberculosis, such as meningitis and miliary, which were nearly always fatal, now have a better than even chance of recovery. Streptomycin pointed a way.'

From Waksman's Nobel banquet speech (10 December 1952)

Born in Priluka, near Kiev in the Russian Empire (now in Ukraine), Waksman moved to the US in 1911 to study agriculture on a scholarship at Rutgers College in New Jersey, where he graduated with a bachelor's degree in 1915. In 1916, he became a naturalized American citizen and he earned a PhD in biochemistry at the University of California in 1918. After that he returned to Rutgers, where he became full professor in 1930 and remained until his retirement in 1958. He also held a post as a marine biologist at the Woods Hole Oceanographic Institution in Massachusetts.

During the 1940s, Waksman and his colleagues isolated several new antibiotics including streptomycin and neomycin. Both of these drugs found extensive application in the treatment of infectious diseases of people, animals and plants. Streptomycin was the first antibiotic that could be used to cure tuberculosis and it saved millions of lives.

For this work, Waksman won the 1952 Nobel prize for physiology or medicine. He was a prolific writer, writing more than 400 scientific papers during his career as well as 28 books including his autobiography *My Life with the Microbes* (1954).

DOROTHY MARY HODGKIN (NÉE CROWFOOT)
12 May 1910 – 29 July 1994

British chemist who developed X-ray crystallography techniques to reveal the structures of important biomolecules

'I became captivated by the edifices chemists had raised through experiment and imagination – but still I had a lurking question. Would it not be better if one could really "see" whether molecules as complicated as the sterols, or strychnine, were just as experiment suggested?'

From Hodgkin's Nobel lecture (11 December 1964)

Hodgkin was born in Cairo, Egypt, where her father was working in the Egyptian Education Service. She studied at the Oxford University then at Cambridge, where she earned her PhD in 1935. Shortly before that she had already returned to Oxford University, where she remained for the rest of her career, becoming full professor in 1960.

Hodgkin began using X-ray crystallography (developed previously by **William** and **Lawrence Bragg**) along with computer analysis to study the structures of complex protein molecules. Her early experiments showed that X-ray exposure can damage crystals, a problem crystallographers still face today. During the Second World War she studied the structure of penicillin, confirming that it has the structure proposed by **Chain**. She also discovered the correct structure for other important molecules including vitamin B12, and for all this work, she won the 1964 Nobel prize for chemistry. She remains the only female British scientist to have won a Nobel prize to date.

After improving X-ray crystallography techniques to study ever more complex molecules, Hodgkin eventually unravelled the structure of the hormone insulin in 1969. She was also a political and peace activist who campaigned vigorously against nuclear weapons.

(JULIUS) ROBERT OPPENHEIMER
22 April 1904 – 18 February 1967

American physicist best known for developing the first atomic weapons

'It is with appreciation and gratefulness that I accept from you this scroll for the Los Alamos Laboratory, and for the men and women whose work and whose hearts have made it. It is our hope that in years to come we may look at the scroll and all that it signifies, with pride. Today that pride must be tempered by a profound concern. If atomic bombs are to be added as new weapons to the arsenals of a warring world, or to the arsenals of the nations preparing for war, then the time will come when mankind will curse the names of Los Alamos and Hiroshima. The people of this world must unite or they will perish.'

From Oppenheimer's acceptance speech for a military award (16 November 1945)

Oppenheimer's father was a wealthy Jewish textile importer who had emigrated from Germany to the US. Oppenheimer studied chemistry at Harvard University, where he earned his first degree in 1925. He was awarded a doctorate just two years later after studying at the universities of Cambridge and Göttingen, Germany, where he met many leading scientists including **Heisenberg**, **Pauli**, **Dirac**, **Fermi** and **Teller**. In 1928, Oppenheimer was appointed as associate professor of physics at the California Institute of Technology in Pasadena and from 1929 to 1942 he worked at the University of California, Berkeley.

His early research was in quantum mechanics. With **Born**, he clarified the physics of the spectra of molecules in quantum terms as well as the interactions between atoms and electrons. He also showed that the proton could not be the equivalent of the anti-electron, or positron, that Dirac had predicted to exist, arguing that the positron's mass would have to be the same as that of an electron while protons are much heavier.

This paved the way for American physicist Carl Anderson to discover the true positron in 1932, for which he received the 1936 Nobel prize for physics. Oppenheimer also worked on cosmic-ray theory and studied reactions of deuterons, the nuclei of a heavy isotope of hydrogen. In other work he developed theories to describe superdense stars such as white dwarfs and neutron stars, and he showed that above a certain mass such stars would become unable to support their own weight and would collapse into what's now known as a black hole.

But Oppenheimer is best known for his high-profile role in directing the development of the atomic bomb at Los Alamos in New Mexico during the Second World War. Oppenheimer set up the laboratory for weapons development and gathered many of the brightest scientists to work on the project. At Los Alamos, he managed more than 3,000 people and tackled the theoretical and mechanical problems that arose.

On 16 July 1945, Oppenheimer witnessed the first atomic bomb explosion in the New Mexico desert. He later said it reminded him of the Hindu scripture, the Bhagavad Gita: 'Now, I am become Death, the destroyer of worlds'. Within a month, atomic bombs were dropped on the Japanese cities Hiroshima and Nagasaki, forcing Japan's surrender.

After the war, when the Manhattan Project was finally public knowledge, Oppenheimer became a famous household name and a national spokesman for science. He briefly returned to the California Institute of Technology before becoming director of the Institute for Advanced Study in Princeton, New Jersey, a post he held until his death. He also chaired the US Atomic Energy Commission. Oppenheimer was initially a vocal opponent of plans to develop an even more powerful hydrogen (fusion) bomb, which was approved nonetheless by President Truman, but he later backed the project once technical hurdles had been ironed out, arguing that Soviet Union would inevitably develop the hydrogen bomb as well.

In 1954, at the height of US anticommunist feeling, Oppenheimer was accused of having communist sympathies and had his security clearance revoked, ending his influence on science policy. **Teller** testified against Oppenheimer, saying that he considered him loyal, but recommending that his clearance be revoked due to poor judgement. Oppenheimer was nominated to receive the Nobel prize for physics three times, but never actually won the award. He was diagnosed with throat cancer in late 1965 and died of the disease in 1967.

ERNST WALTER MAYR
5 July 1904 – 3 February 2005

German-born American biologist whose work led to the modern evolutionary synthesis of Mendelian genetics and Darwinian evolution

'The idea that a few people have about the gene being the target of selection is completely impractical; a gene is never visible to natural selection, and in the genotype, it is always in the context with other genes, and the interaction with those other genes make a particular gene either more favourable or less favourable ... Now we know that it is really the whole genotype of the individual, not the gene. Except for that slight revision, the basic Darwinian theory hasn't changed in the last 50 years.'

From an interview with Mayr at www.edge.org

Born in Kempten, Bavaria, Mayr developed an early interest in ornithology and studied medicine at the University of Greifswald from 1923. He earned a doctorate in ornithology from the University of Berlin in 1926 and then worked at the Berlin Museum before emigrating to the US, where he accepted an invitation to join the staff of the American Museum of Natural History in New York in 1931. He was professor at Harvard University from 1953 until his retirement in 1975.

Early in his career, Mayr undertook an expedition to the Dutch East Indies (now Indonesia), where he identified many new animal and plant species unknown in the west. In his classic book *Systematics and the Origin of Species* (1942), Mayr revised the concept of a species, previously defined in terms of a group of individual creatures with common features. Mayr introduced the modern view of a species as a group whose members can interbreed only with one another, and this led to a clearer view of how species diverge.

He was also first to fully describe the 'founder effect', which occurs when a small number of individuals become isolated from the main population, for instance when a small group newly colonizes an island. The founding group rapidly consolidates genetic change and can form a new species because they have much less genetic variation than their ancestral group.

HANS ADOLF KREBS
25 August 1900 – 22 November 1981

German-born British biochemist who identified the key chemical reactions that generate energy in cells

'The research I have been doing – studying how foodstuffs yield energy in living cells – does not lead to the kind of knowledge that can be expected to give immediate practical benefits to mankind ... I am convinced that an understanding of the process of energy production will eventually help us in solving some of the practical problems of medicine.'

From Krebs's Nobel banquet speech (10 December 1953)

Born to a Jewish family in Hildesheim, Germany, Krebs studied medicine at the universities of Göttingen, Freiburg-im-Breisgau and Berlin between 1918 and 1923. In 1925, he earned his MD degree at the University of Hamburg then he spent a year studying chemistry in Berlin. He later worked at the Kaiser Wilhelm Institute for Biology at Berlin-Dahlem and in hospitals until the Nazi government terminated his appointment in 1933. He then emigrated to England, where he later became professor at the universities of Sheffield and Oxford.

Krebs's wide-ranging research included investigations of how urea is synthesized in the mammalian liver and how the body oxidizes foodstuffs. He is best known for explaining what's now called the Krebs cycle (or citric acid cycle), the chemical steps involved when food is broken down to provide the cells of all aerobic living organisms with high-energy adenosine triphosphate (ATP) molecules. This molecule supplies the energy to build up proteins from amino acids and replicate DNA as cells divide.

For this work, Krebs shared the 1953 Nobel prize for physiology or medicine with German–American biochemist Fritz Lipmann. He was knighted in 1958.

OSWALD THEODORE AVERY
21 OCTOBER 1877 – 2 FEBRUARY 1955

Canadian-born American physician who discovered that DNA is the carrier of genetic information

'If we are right, and of course that is not yet proven, then it means that nucleic acids are not merely structurally important but functionally active substances in determining the biochemical activities and specific characteristics of cells and that by means of a known chemical substance it is possible to induce predictable and hereditary changes in cells. This is something that has long been the dreams of geneticists.'

From a letter Avery sent to his brother Roy (1943)

From Halifax, Nova Scotia, Avery moved to Hamilton, New York to study at Colgate University. He then spent most of the rest of his life in New York City. He earned his MD degree in 1904 from Columbia University College of Physicians and Surgeons and then practised medicine until 1907 when he took a research post at the Hoagland Laboratory in Brooklyn. In 1913, he moved to New York's Rockefeller Institute for Medical Research (now the Rockefeller University).

Before the 1940s, the chemical identity of genes was completely unknown and many scientists assumed genetic information was carried by proteins, which seemed to be the only biochemicals that exist in the vast array of forms thought necessary to store all of an organism's complex genetic information. It was widely assumed that the four chemical bases in DNA (deoxyribonucleic acid) repeated without variation in the chromosomes of all species. An intriguing experiment by Avery and his colleagues proved that DNA nonetheless does carry our genetic information while the team was following up work by a British microbiologist, Frederick Griffith.

Griffith had studied two strains of the bacterium *Streptococcus pneumonia*, which can cause pneumonia. He noticed that they varied dramatically in appearance and virulence, or their ability to cause disease.

The highly virulent S strain had a smooth outer capsule, while the benign R strain looked rough and had no capsule. Mice injected with the S strain died within a few days of injection, while mice injected with the R strain did not die.

In experiments, Griffith established that the virulence of the S strain was destroyed by heating the bacteria. So it puzzled him that when he injected mice with both heat-killed S bacteria and living R bacteria – neither of which caused disease on their own – the mice died. Isolating living bacteria from the dead mice, he found that they now had the smooth capsules characteristic of the S strain. Somehow, a chemical component from the heat-killed S cells had transformed the living R cells into the virulent S form.

Hearing of these results, Avery and his colleagues Colin MacLeod and Maclyn McCarty cultured the R and S strains to try to identify the specific molecules in the heat-treated S strain that transform living R cells into the virulent S strain. They used enzymes to selectively destroy protein, RNA or DNA in heat-treated S cells and then mixed them with living R cells. The R strain always turned into the S strain unless the S strain's DNA had been destroyed, proving that DNA was the molecule that caused the crucial transformation.

In 1944, the team published their results in *The Journal of Experimental Medicine* (by now Avery was in his late sixties). The experimental findings of the Avery–MacLeod–McCarty experiment were quickly replicated and also extended to other hereditary characteristics. However, many scientists were still reluctant to accept the conclusion that DNA was the genetic material. But later work by **Hershey** confirmed DNA as the carrier of genetic information, while American biochemist Erwin Chargaff reported in 1950 that while some kind of pairing occurs in DNA bases (the number of adenine bases is always equal to the number of thymine bases, while the number of cytosine bases is always equal to the number of guanine bases), the composition of DNA does vary between species.

This evidence of molecular diversity made DNA a much more plausible candidate for a carrier of genetic information and paved the way for the discovery of the helical structure of DNA by **Crick** and **Watson**, as well as a clear understanding of how genes are encoded by sequences of base pairs. **Nirenberg** and others cracked the genetic code in the early 1960s, by explaining how DNA is translated into proteins.

NORMAN ERNEST BORLAUG
25 March 1914 – 12 September 2009

American agronomist and a central figure in the 'green revolution'

'There can be no permanent progress in the battle against hunger until the agencies that fight for increased food production and those that fight for population control unite in a common effort. Fighting alone, they may win temporary skirmishes, but united they can win a decisive and lasting victory to provide food and other amenities of a progressive civilization for the benefit of all mankind.'

From Borlaug's Nobel prize acceptance speech (10 December 1970)

Borlaug was born near Cresco in Iowa and studied forestry at the University of Minnesota. After that he worked for the US Forestry Service at stations in Massachusetts and Idaho and then he returned to the University of Minnesota to study plant pathology, earning his doctorate there in 1942. From 1984, he taught and researched at Texas A&M University until his death.

In 1944, Borlaug was tasked with directing the Cooperative Wheat Research and Production Program in Mexico. Through scientific research in plant genetics and breeding, he developed extremely high-yielding, disease-resistant strains of wheat. As a result, Mexico became a net exporter of wheat by 1963. During the 1960s, he also collaborated with scientists from other parts of the world, especially from India and Pakistan, to increase wheat yields in new climates.

Borlaug attracted criticism for not being green in the modern sense of the word; he endorsed artificial fertilizers and chemical pesticides for high crop yields, and he became a vocal backer of genetically modified crops. But he is often credited with saving the lives of more than a billion people who would otherwise have died in famines. For his contributions to the world food supply, Borlaug won the 1970 Nobel prize for peace.

GEORGE GAYLORD SIMPSON
16 June 1902 – 6 October 1984

American palaeontologist who reconciled Darwinian evolution by natural selection with the fossil record

'Man is the result of a purposeless and natural process that did not have him in mind'

From Simpson's *The Meaning of Evolution* (1967 edition)

Simpson studied at the University of Colorado and at Yale University in Connecticut, where he earned a PhD in 1926. In 1927, he joined the staff of the American Museum of Natural History in New York, and from 1959 to 1970 he was professor at Harvard. Later he worked at the University of Arizona until his retirement in 1982.

Simpson was among the first scientists to use mathematical methods in palaeontology. His key contribution was to iron out disputes over whether natural populations contained enough genetic variation for Darwinian natural selection to create new species. Many scientists thought other effects must take place, such as inheritance of acquired characteristics, or some mysterious goal-orientated 'directed variation'.

Simpson showed from the fossil record that there was no need to resort to these processes to explain the emergence of new species. While some scientists assumed modern horses had evolved their current features in a linear process of change, Simpson recognized that its evolution had progressed more like an irregular tree with many side branches that ended in extinction. He also anticipated the concept of punctuated equilibrium, a theory developed by **Gould** which holds that evolutionary change happens in fits and starts, sometimes moving very fast, but with long periods of slow change or none at all.

He also showed that the fossil record fitted perfectly with natural selection mechanisms if evolution acts on gene pools in populations, not on individual members. He explained some gaps in the fossil record as due to rapid changes in small populations that left little fossil evidence to document them. Simpson wrote more than a dozen books, including the classic *Tempo and Mode in Evolution* (1944) and *The Meaning of Evolution* (1949), which was widely sold and translated into 10 languages.

GERTRUDE BELLE ELION
23 January 1918 – 21 February 1999

American pharmacologist who developed life-saving drugs including chemotherapy

'I was very close to [my grandfather] because he came over
from Europe when I was about three years old and lived
very close to us ... Also, I watched him die, essentially, in the
hospital. And that made a terrific impression on me. I decided
that nobody should suffer that much.'

Elion recalls that her grandfather's death from cancer may have been
a deciding factor in her career (from an interview for the
American Academy of Achievement, 6 March 1991)

Born in New York City, Elion studied science at Hunter College
in Manhattan from 1933. After working as an assistant in a chemistry
lab, she earned an MSc in chemistry from New York University in 1941.
After various short research posts, she became assistant to George
Hitchings at the Burroughs-Wellcome pharmaceutical company (now
GlaxoSmithKline), where she remained for the rest of her career.

Until the 1940s, drug development had largely focused on chemical
modification of natural products, then testing by trial and error. Elion
and Hitchings took a new approach. By painstakingly understanding the
biochemical differences between human cells and pathogens (such as
bacteria or viruses) and tumour cells, they developed a host of drugs to
treat disease with minimum harm to normal healthy cells. They introduced
the first successful treatment for acute leukaemia, as well as acyclovir (for
herpes), the first effective antiviral drug. They also developed treatments
for malaria and meningitis as well as azathioprine, an immunosup-
pressant drug that prevents rejection of kidney transplants and alleviates
rheumatoid arthritis.

For their work on drug discovery, Elion and Hitchings shared the
1988 Nobel prize for physiology or medicine with Scottish pharmacologist
James Black, who developed beta-blockers to treat heart disease.

BARBARA McCLINTOCK
16 June 1902 – 2 September 1992

American geneticist who discovered 'jumping genes'

'The maize plant, with which I have worked for many years,
revealed a genetic phenomenon that was totally at odds with
the dogma of the times, the mid-nineteen forties ...
I was not invited to give lectures or seminars, except on
rare occasions ... Instead of causing personal difficulties,
this long interval proved to be a delight. It allowed complete
freedom to continue investigations without interruption, and
for the pure joy they provided.'

From McClintock's Nobel banquet speech (10 December 1983)

McClintock studied genetics at Cornell University, where she earned her first degree in 1923 and a PhD in 1927. After taking several posts at Cornell and elsewhere, she joined the staff of the Carnegie Institution of Washington in Cold Spring Harbor, New York, in 1942 and remained there until 1967.

McClintock was famous for her 1940s discovery of 'transposable elements', also known as jumping genes. These are sequences of DNA that can hop from one position in a cell's genome to another position, either via 'copying and pasting' or 'cutting and pasting'. McClintock discovered the first jumping genes in maize and noticed that they were capable of insertions, deletions and translocations that could lead to physical changes in the plants, such as a colour change of the corn kernels.

For this work, she won the 1983 Nobel prize for physiology or medicine. Today transposable elements are known to play key roles in disease by damaging the genome of host cells, but they can also confer benefits in evolution by allowing the creation of new genes. They have also proved useful in experiments where researchers deliberately want to induce genetic mutations.

English astronomer and mathematician who made key discoveries about energy generation in stars

'A junkyard contains all the bits and pieces of a Boeing 747, dismembered and in disarray. A whirlwind happens to blow through the yard. What is the chance that after its passage a fully assembled 747, ready to fly, will be found standing there? So small as to be negligible, even if a tornado were to blow through enough junkyards to fill the whole Universe.'

Hoyle expresses how unlikely he thought it that even simple life could chemically evolve on Earth (*The Intelligent Universe*, 1983)

Hoyle was born near Bingley in West Yorkshire. He studied maths at Cambridge University and remained there for most of his career, becoming Plumian Professor of Astronomy and Experimental Philosophy in 1958. From 1972 to 1978, he was professor at Cornell University in Ithaca, New York.

In 1946, Hoyle established the concept of nucleosynthesis in stars. This explained how heavy elements like carbon and oxygen were built up in the Universe gradually inside stellar cores. As stars eject their atmospheres at the ends of their lives or explode in supernovae, these heavy elements are recycled into new stars that form from interstellar clouds, each generation of stars building up yet heavier elements. Hoyle also proposed correctly that some rare heavy elements could form inside supernovae with their vast temperatures and pressures.

But Hoyle famously rejected the Big Bang theory, a term he himself coined. He was troubled by the idea that the Universe had a fixed origin in time, suggesting a creator or at least some kind of mystifying trigger. Instead he backed the 'steady-state' theory of the Universe. He accepted **Hubble**'s observations suggesting the Universe is expanding, but proposed a 'continuous creation' model in which matter spontaneously fills space between galaxies over time, so that even though galaxies get farther apart, new ones that develop between them fill the space they leave.

Hoyle was quick to criticize many conventional theories, even when they had widespread acceptance. As late as the 1980s, he argued that oil had a primordial origin rather than a biological one, stating: 'The suggestion that petroleum might have arisen from some transformation of squashed fish or biological detritus is surely the silliest notion to have been entertained by substantial numbers of persons over an extended period of time.'

Late in life, Hoyle became an ardent proponent of panspermia, the idea that life first evolved not on Earth but in space. With his Sri Lankan colleague Chandra Wickramasinghe at the University of Cardiff, he developed the theory that for billions of years, microbes have been ferried around the solar system, and possibly beyond, inside bodies like comets as well as interstellar clouds, and this extraterrestrial life seeded the Earth with life forms. The theory could explain why life on Earth emerged surprisingly soon after the hot, young planet became habitable, but there is no direct evidence for it.

Hoyle was knighted in 1972. He became famous for his popular science communication, as a broadcaster and author, his books including *The Intelligent Universe* (1983). He also prolifically wrote science fiction. His first novel *The Black Cloud* describes most intelligent life in the Universe as interstellar gas clouds, which are surprised to discover that intelligent civilizations also thrive on planets.

MAX DELBRÜCK
4 September 1906 – 9 March 1981

German–American biophysicist who pioneered molecular biology and virus research

'Molecular genetics, our latest wonder, has taught us to spell out the connectivity of the tree of life in such palpable detail that we may say in plain words: this riddle of life has been solved'

From Delbrück's Nobel lecture (10 December 1969)

Delbrück was born in Berlin, Germany; his mother was a granddaughter of **Liebig**. He studied astrophysics and later theoretical physics at Göttingen, where he earned his PhD in 1929, then spent three postdoctoral years in England, Switzerland and Denmark. In Copenhagen, he worked with **Bohr**, who inspired him to consider interesting connections between physics and biology. In 1932, he became an assistant to **Meitner** in Berlin and then joined a Berlin research group that used radiation to produce mutations in fruit flies. In 1937 he emigrated to the US, where he worked at the California Institute of Technology for most of his career.

Delbrück is best known for his work on bacteriophage, a type of virus that infects bacteria rather than animal or plant cells. Working with Salvador Luria, Delbrück devised an important equation describing the mutation rate of bacteria from virus-susceptible to virus-resistant forms. In experiments, he showed independently of **Hershey** that two different viruses could recombine inside a bacterial cell, exchanging sections of their genetic material.

For their studies of the genetic structure of viruses, Delbrück, Hershey and Luria shared the 1969 Nobel prize for physiology or medicine. **Schrödinger**'s book *What Is Life?* (1944) popularized Delbrück's views about the physical basis of life, speculating about what genes are and how they might behave – the book impressed **Crick**, who with **Watson** would later discover the structure of DNA.

EDWARD MILLS PURCELL
30 August 1912 – 7 March 1997

American physicist who discovered nuclear magnetic resonance in liquids and solids

'One can detect the precession of the magnetic nuclei in a drop of water... I have not yet lost a feeling of wonder, and of delight, that this delicate motion should reside in all the ordinary things around us, revealing itself only to him who looks for it. I remember, in the winter of our first experiments, just seven years ago, looking on snow with new eyes. There the snow lay around my doorstep – great heaps of protons quietly precessing in the Earth's magnetic field.'

From Purcell's Nobel lecture (11 December 1952)

Purcell studied electrical engineering at Purdue University in Indiana, graduating in 1933. He earned a PhD in physics from Harvard University in 1938. After serving two years as instructor in physics at Harvard, he joined the Massachusetts Institute of Technology's Radiation Laboratory, founded in 1940 for military research and development of microwave radar. After the Second World War he returned to Harvard, where he was appointed full professor in 1949 and remained for the rest of his career.

During the war, Purcell worked with **Rabi**, who had first described and measured nuclear magnetic resonance in molecular beams in 1938. In 1946, Purcell published a report describing his discovery, independently of **Bloch**, of nuclear magnetic resonance in liquids and solids. Nuclear magnetic resonance gave scientists their most precise way of examining the structure of materials and also led to magnetic resonance imaging, which has radically improved diagnosis of disease.

For his work on nuclear magnetic resonance, Purcell shared with Bloch the 1952 Nobel prize for physics. In radio astronomy, he made the first detection of a key wavelength (21 cm) emitted by neutral hydrogen in the Milky Way. In the 1950s, radio telescopes tuned to this wavelength revealed the spiral arms of our galaxy for the first time. Purcell also served as a science adviser to three US presidents.

JOHN BARDEEN
23 May 1908 – 30 January 1991

American physicist who co-invented the transistor and explained superconductivity

'Many years of research by a great many people, both before and after the discovery of the transistor effect, has been required to bring our knowledge of semiconductors to its present development. We were fortunate enough to be involved at a particularly opportune time and to add another small step in the control of Nature for the benefit of mankind.'

From John Bardeen's Nobel lecture (11 December 1956)

Born in Madison, Wisconsin, Bardeen studied electrical engineering at the University of Wisconsin, graduating in 1928. He stayed there as a research assistant, working on mathematical problems in applied geophysics and radiation from antennas. During 1930 to 1933, he worked at the Gulf Research Laboratories in Pittsburgh, Pennsylvania, on the interpretation of magnetic and gravitational surveys for oil prospecting. He then did a PhD in mathematical physics at Princeton University in New Jersey, which he was awarded in 1936.

During the Second World War, Bardeen conducted research relating to underwater ordnance and mine sweeping, and after that joined Bell Telephone Laboratories, where he remained until 1951, when he was appointed professor at the University of Illinois. While at Bell labs, Bardeen, **Shockley** and Walter Brattain developed the point-contact transistor – the first type of solid-state electronic transistor – in 1947. It dramatically reduced the size of computer technology, which had previously relied on bulky vacuum tubes as electrical switches. For this work, Bardeen, Shockley and Brattain shared the 1956 Nobel prize for physics.

Bardeen won a Nobel prize again in 1972 (with Leon Cooper and John Schrieffer) for proposing the first successful explanation of superconductivity, in which materials lose all their electrical resistance below a characteristic temperature. Bardeen was one of only four people to have won a Nobel prize twice.

WILLIAM BRADFORD SHOCKLEY
13 February 1910 – 12 August 1989

American physicist who was one of the inventors of the transistor

'I am overwhelmed by an irresistible temptation to do my climb by moonlight and unroped. This is contrary to all my rock climbing teaching and does not mean poor training, but only a strong-headedness.'

Shockley's 1947 memo to himself regarding work on the transistor
(from *Broken Genius* by Joel Shurkin, 2006)

Shockley grew up in Palo Alto, California. He studied at the California Institute of Technology, graduating with a first degree in 1932, and then earned a PhD at the Massachusetts Institute of Technology in 1936. That year he joined Bell Telephone Laboratories.

During the Second World War he worked on antisubmarine warfare operations research and gave advice to the US War Department about the potentially enormous casualties of a Japanese invasion, which influenced the decision to drop atom bombs on Hiroshima and Nagasaki to force a Japanese surrender. He returned to Bell Labs after the war until 1955. Working there with his colleagues **Bardeen** and Walter Brattain, Shockley developed semiconductor transistors (used to amplify and switch electronic signals in computers) to replace the bulky thermionic valves used previously. This played a key role in the miniaturization of circuits in computers.

For this work, Shockley, Bardeen and Brattain shared the 1956 Nobel prize for physics. From the mid-1950s, Shockley ran his own company in Mountain View, California, to commercialize semiconductor devices, and in 1963 he was appointed professor at Stanford University. A controversial figure, he was notorious for his abrasive management style and support of unpopular theories of inherited intelligence among different races.

RALPH ASHER ALPHER
3 February 1921 – 12 August 2007

American physicist who predicted the existence of the cosmic microwave background, the relic radiation of the Big Bang that created the Universe

'Four Nobel prizes have been given for measurements of the cosmic background radiation. Everyone familiar with the history of cosmology realizes that Alpher and Herman, who predicted its existence and temperature in advance, should have been so recognized too. Whatever the Swedish Academy thinks, they are Nobel Laureates in the minds of the community whose opinion matters.'

Dwight Neuenschwander, editor of the American Institute of Physics journal *Radiations*, mourns the failure of Alpher to be awarded a Nobel prize (spring 2009)

Alpher left school aged 15 to earn money for his family doing administrative work. At the same time he studyied physics at the night school of George Washington University, where he earned a first degree in 1943 and a PhD in 1948. During the Second World War, he also worked on secret defence projects.

Alpher's thesis adviser, **Gamow**, suggested that he investigate the very beginning of time. The idea that the Universe came into being in a hot Big Bang explosion had been controversially proposed about 25 years earlier. Alpher's PhD theoretically described the production of elements during the first minutes after the Big Bang, revealing that 90 per cent of hydrogen and helium in the Universe was produced in the first 17 seconds.

Soon after, Alpher and his colleague Robert Herman predicted the existence of the cosmic microwave background, the 'afterglow' of the Big Bang. This radiation would have flooded the Universe and been stretched to microwave wavelengths by the expansion of the Universe. Detection and studies of this radiation later earned Nobel prizes for **Penzias**, **Wilson**, **Smoot** and **Mather**.

HENDRIK BRUGT GERHARD CASIMIR
15 July 1909 – 4 May 2000

Dutch physicist who predicted a strange attractive force now called the Casimir effect

'During my lifetime, the attitude of scientists towards their science and the attitude of society towards scientists has changed. Scientists no longer feel safe and smugly virtuous in their ivory towers and society ... tends to regard them as potentially dangerous.'

From Casimir's memoirs *Haphazard Reality* (1983)

Casimir was born in The Hague and studied theoretical physics at the University of Leiden, where he earned his PhD in 1931. During that time he also worked with **Bohr** in Copenhagen, and afterwards he worked with **Pauli** in Zurich. He was appointed professor of physics at Leiden University in 1938. From the Second World War onwards, he worked in industry at Philips laboratories in Eindhoven but remained active in scientific research.

Casimir made key contributions to theories of superconductivity, the disappearance of electrical resistance in materials. But he is best known for predicting the existence of the 'Casimir effect', a tiny attractive force acting between two parallel, uncharged conducting plates in a vacuum. This attraction arises because a vacuum is not just empty space, but is seething with energy, with particles constantly fleeting in and out of existence.

In 1948, Casimir pointed out that close metal plates would block out light waves that are too big to fit between them. If the gap was only a few nanometres (billionths of a metre) wide, the energy density outside the plates would become significantly higher than between the plates, creating a pressure that pushes the plates together.

The Casimir effect was first accurately measured in 1997, and can also be a repulsive force, depending on the experimental arrangement. It could one day be useful in nanoscale machinery, creating repulsion between tiny components that allows them to move without friction.

FREEMAN JOHN DYSON
15 DECEMBER 1923 –

British–American physicist renowned for futuristic speculations

'The biggest breakthrough in the next 50 years will be the discovery of extraterrestrial life. We have been searching for it for 50 years and found nothing. That proves life is rarer than we hoped, but does not prove that the universe is lifeless. We are only now developing the tools to make our searches efficient and far-reaching, as optical and radio detection and data processing move forward. I consider it likely that evidence of extraterrestrial life will be found before 2056.'

Dyson forecasts the future (*New Scientist* online, 16 November 2006)

As a young man during the Second World War, Dyson joined the operational research section at RAF Bomber Command, and afterwards did a maths degree at Cambridge University. In 1947 he moved to the US to take up a fellowship at Cornell University, where he later became professor. In 1953, he moved to Institute for Advanced Study in Princeton, New Jersey, where he has remained for the rest of his career. He became a naturalized American citizen in 1957.

At Cornell, Dyson made an important contribution to quantum electrodynamics (the theory of how particles and light interact) by showing that different theories in use at the time were mathematically equivalent. His work and lectures clarified **Feynman**'s quantum electrodynamics theories, leading to greater acceptance of Feynman's work in the scientific community. He also did wide-ranging work on maths topics including number theory and topology, the study of the continuity and connectivity of abstract objects.

Dyson was also one of the leaders of the Orion project to design spacecraft propulsion by a series of nuclear explosions, although the project was shelved following the Partial Test Ban Treaty of 1963. He also led the design of TRIGA, a small nuclear reactor that can be installed without a containment building and still used by scientific institutions for education and research. They have many diverse applications, including production of radioisotopes for medicine and industry, as well as treatment of tumours.

In popular culture he is most famous for his imaginative futuristic speculations on a wide range of topics including extraterrestrial intelligence. In 1960, he proposed the concept of the Dyson sphere, a hypothetical shroud of solar power satellites that an alien civilization would use to capture all the energy output of their planet's parent star. He deemed this the logical consequence for any thriving technological civilization that needs to maximize its energy reserves and proposed that searching for these structures might be the best strategy in the search for advanced civilizations in the Universe.

He is also renowned for his questioning stance about global warming, claiming that predictions of temperature increases and their consequences have been 'grossly exaggerated'. In August 2007, he wrote in the online magazine Edge.org: 'I am opposing the holy brotherhood of climate model experts and the crowd of deluded citizens who believe the numbers predicted by the computer models ... They do not begin to describe the real world that we live in. The real world is muddy and messy and full of things that we do not yet understand.'

Dyson has written extensively about the relationship between science and religion. His many books include the autobiographical *Disturbing the Universe* (1979) and *The Sun, the Genome and the Internet* (1999), which envisions green technologies enabling rural communities to thrive, halting the migration from villages to vast cities.

RICHARD PHILLIPS FEYNMAN
11 May 1918 – 15 February 1988

Colourful American physicist who developed quantum electrodynamics

'Poets say science takes away from the beauty of the stars – mere globs of gas atoms. Nothing is "mere". I too can see the stars on a desert night, and feel them ... It does not do harm to the mystery to know a little more about it. For far more marvellous is the truth than any artists of the past imagined it. Why do the poets of the present not speak of it? What men are poets who can speak of Jupiter if he were a man, but if he is an immense spinning sphere of methane and ammonia must be silent?'

From *The Feynman Lectures on Physics* (1964)

Feynman developed an early interest in science growing up in New York City, and set up a home laboratory where he experimented with electricity, inventing a burglar alarm and repairing damaged circuits in radios. He studied at the Massachusetts Institute of Technology and at Princeton University in New Jersey, where he earned a PhD in 1942.

During the Second World War, Feynman worked on the Manhattan Project to develop atomic weapons. His role involved developing methods to separate the two isotopes uranium-235 and uranium-238. While working at Los Alamos, he quickly gained a reputation as a practical joker and he was fond of highlighting security lapses by cracking the safes in which designs for bombs were stored. From 1945 to 1950, he was professor of theoretical physics at Cornell University and after that he held several professorships at the California Institute of Technology.

In the late 1940s, Feynman developed the theory of quantum electro-dynamics, which describes how charged particles interact by exchanging photons of light. For the first time, this description was consistent with both quantum mechanics and **Einstein**'s theory of special relativity. For his work on quantum electrodynamics, Feynman shared the 1965 Nobel prize for physics with the Japanese physicist Shin-Ichiro Tomonaga and his fellow American Julian Schwinger, who both made contributions to the same theory.

Feynman also correctly proposed that hadrons (including protons and neutrons) are composed of discrete point-like 'partons', now commonly referred to as quarks and gluons, and he introduced 'Feynman diagrams', simple graphs for visualizing particle interactions. He was a prolific and entertaining writer who became famous for titles such as *Surely You're Joking, Mr Feynman!* (1985) and *What Do You Care What Other People Think? Further Adventures of a Curious Character* (1988).

Feynman had a passion for promoting excellent education. During a visit to Brazil, he evaluated the Brazilian educational system and he was also involved in overseeing maths and physics texts for primary and secondary public schools in California. His lectures at the California Institute of Technology were published in the three-volume book *The Feynman Lectures on Physics* (1964), which has inspired physics students ever since. Estimates suggest this book has sold more than 1.5 million copies in English as well as a similar number in other languages. During his lifetime, Feynman was one of the most famous scientists in the world.

Late in his life, Feynman was appointed as a member of the committee set up to investigate the cause of the Challenger space shuttle disaster in 1986, when seven astronauts were killed because the shuttle broke apart 73 seconds into its flight. He was deeply critical of NASA and of the arguments used to justify the shuttle's safety, writing: 'There are several references to flights that had gone before. The acceptance and success of these flights is taken as evidence of safety. But erosion and blow-by are not what the design expected ... The fact that this danger did not lead to a catastrophe before is no guarantee that it will not the next time, unless it is completely understood. When playing Russian roulette the fact that the first shot got off safely is little comfort for the next.'

Though raised Jewish, Feynman was an atheist as an adult and even avoided being described as having Jewish ethnicity. He asked to be omitted from books or lists of famous Jewish people, worrying that the suggestion that certain characteristics stem from Jewish heredity leads to nonsensical racial theories. He also said that as a teenager he stopped believing that Jewish people are in any way 'the chosen people'. He died of abdominal cancer in 1988 aged 69.

GEORGE GAMOW
4 March 1904 – 19 August 1968

Russian-born physicist who made pioneering discoveries in radioactivity and cosmology

'It took less than an hour to make the atoms, a few hundred million years to make the stars and planets, but five billion years to make man!'

From Gamow's *The Creation of the Universe* (1952)

Gamow was born Georgiy Antonovich Gamov in Odessa, Russian Empire (now in Ukraine), and studied at the University of Leningrad from 1923 until 1929. He then worked at various universities in Europe including the University of Copenhagen and Cambridge University, where he collaborated with **Rutherford**. In 1934, he moved the US, where he taught at the George Washington University in Washington DC. He moved to the University of Colorado in 1956.

In his work on radioactivity, Gamow showed that alpha decay (release of an alpha particle containing two neutrons and two protons) occurs through a 'quantum tunnelling' process. Classically, an alpha particle is too tightly bound to an atomic nucleus to escape. However, probabilistic quantum theory allows the particle a small chance of tunnelling out through the potential barrier. Gamow's work explained the relationship between the half-life of an element that undergoes alpha decay and the energy of the particles emitted.

From the late 1930s, Gamow turned his focus to astrophysics and cosmology. With **Alpher**, he showed how reactions following the creation of the Universe in the Big Bang explain the current abundances of hydrogen and helium in the Universe. He even made forays into biology, suggesting correctly in 1954 that triplets of nucleic acids act as a genetic code (although he assumed incorrectly that the triplets can overlap; see **Brenner**). Gamow also wrote popular science books including the *Mr Tompkins* series, in which the main character witnesses strange realms of physics beyond everyday experience.

ALFRED CHARLES KINSEY
23 June 1894 – 25 August 1956

American biologist who pioneered the scientific study of human sexuality

'Males do not represent two discrete populations, heterosexual and homosexual. The world is not to be divided into sheep and goats. Not all things are black nor all things white ... Only the human mind invents categories and tries to force facts into separated pigeon-holes.'

From Kinsey's *Sexual Behavior in the Human Male* (1948)

Kinsey studied engineering at Stevens Institute of Technology in his home town before enrolling at Bowdoin College in Maine in 1914 to study biology and psychology. After graduating there he earned a doctorate at Harvard University in 1919 for work on gall wasps, in which he travelled widely to gather millions of the insects. He became professor of zoology at Indiana University, Bloomington, in 1920.

Kinsey is best known as the founder of modern sexology, the rigorous scientific study of human sexuality. He developed a scale running from 0 to 6 to quantify sexual orientation, in which 0 is exclusively heterosexual and 6 is purely homosexual. He continued his sexology studies at Indiana University's Institute for Sex Research, which he founded in 1942, and this led to the publication of the two Kinsey Reports, one on male sexuality and one on female sexuality, in 1948 and 1953.

The reports detailed the prevalence of homosexuality in men and women and showed that male and female sexual practices are more similar than traditional views suggested. The Kinsey Reports were translated into 13 languages and sold more than three quarters of a million copies, making Kinsey world famous. His work was controversial, however. He was accused of biased results and criticized for observing sexual activity and engaging in sexual experiments with colleagues.

LOUIS SEYMOUR BAZETT LEAKEY
7 August 1903 – 1 October 1972

British anthropologist who clarified the evolution of humans in Africa

'I feel that, perhaps, I know the Kikuyu better than any white man living – I am in so many ways a Kikuyu myself – and that is why I have dared to write this book, hoping and praying that it may help to bring understanding of the problems of the tribe and, in due course, peace to all the people of Kenya.'

From Leakey's *Mau Mau and the Kikuyu* (1952)

Born to missionary parents in Kabete, British East Africa (now Kenya), Leakey grew up with the Kikuyu tribe and then studied archaeology and anthropology at Cambridge University, where he graduated in 1926. He earned a PhD in 1930 for studies of African pre-history. From 1945 to 1961, he was curator of the Coryndon Memorial Museum in Nairobi.

When Leakey began his research, it was commonly believed that humans had originated in Asia. Leakey overturned this view with several fossil discoveries in Africa, starting with the skull of a Miocene hominoid in 1948. This ape-like creature, which Louis named *Proconsul africanus*, lived from about 23 to 14 million years ago and was probably a common ancestor of both humans and other primate species.

Many of Leakey's famous discoveries were found at Olduvai Gorge in northern Tanzania. With his wife Mary Leakey, between 1962 and 1964 he discovered fossils that led him to announce the new species *Homo habilis* ('handy-man'). *H. habilis* lived from about 2.3 to 1.4 million years ago; debates continue as to whether or not it is a direct ancestor of humans.

(FRANK) MACFARLANE BURNET

3 September 1899 – 31 August 1985

Australian virologist who proposed the concept of acquired immunity

'How can an immunized animal recognize the difference
between an injected material like insulin or serum albumin
from another species and its own corresponding substance? ...
Their recognition in the sense in which we are using the word
requires that there be available in the body a large volume of
accessible "information" with some superficial analogies to a
dictionary. In other words, there must be something against
which a configuration can be compared and a decision made
whether it corresponds or not.'

From Burnet's Nobel lecture (12 December 1960)

Born in Traralgon, Victoria, Burnet studied medicine at the
University of Melbourne, where he graduated MD in 1923. He remained
at Melbourne University to research the immune response to typhoid
fever. After a brief spell in London, he returned to Melbourne University
where he spent the rest of his career, becoming professor of experimental
medicine there in 1944.

Burnet's key work was on immunology. After developing techniques to
multiply viruses in chicken embryos, he noticed that embryos don't produce
the appropriate antibodies to combat the viral infection. From this he came to
recognize the concept of acquired immunity – if an embryo can't distinguish
its own tissue from foreign tissue, then that ability must be 'learned' as the
immune system encounters foreign substances (antigens), attacks them and
remembers them to thwart future invasions much more quickly.

The British biologist Peter Medawar went on to prove experimentally
that an embryo can indeed develop immunological tolerance through
exposure to antigens. For their work on acquired immunity, Burnet and
Medawar shared the 1960 Nobel prize for physiology or medicine. Burnet
was knighted in 1951.

American physical chemist who developed radiocarbon dating

'The dating technique itself is one which requires care, but which can be carried out by adequately trained personnel who are sufficiently serious-minded about it. It is something like the discipline of surgery – cleanliness, care, seriousness, and practice. With these things it is possible to obtain radiocarbon dates which are consistent and which may indeed help roll back the pages of history and reveal to mankind something more about his ancestors.'

From Libby's Nobel lecture (12 December 1960)

 **Born in Grand Valley, Colorado**, Libby studied chemistry at the University of California, Berkeley, where he earned a first degree in 1931 and a PhD in 1933. He remained there as a lecturer and then assistant professor before joining the Manhattan Project to build atomic weapons during the Second World War. In 1945 he became professor at the University of Chicago and from 1959 he was professor at the University of California, Los Angeles.

In 1949, Libby led a team that introduced the technique of radiocarbon dating, a powerful method for estimating the ages of carbon-bearing materials up to about 60,000 years old. It relies on measuring the content of carbon-14, an isotope that decays with a half-life of about 5,700 years. Constant interchange with the environment makes the ratio of carbon-14 to carbon-12 constant in a living tree, for instance, but the ratio drops with time in a predictable way after the tree dies.

Libby demonstrated the accuracy of radiocarbon dating by using it to accurately measure the ages of wood samples with known ages, including an ancient Egyptian barge built around 1850 BC. For introducing the technique, Libby won the 1960 Nobel prize for chemistry.

ROSALYN YALOW (NÉE SUSSMAN)
19 July 1921 – 30 May 2011

American medical physicist who developed the technique of radioimmunoassay

'The first telescope opened the heavens; the first microscope opened the world of the microbes; radioisotopic methodology, as exemplified by radioimmunoassay, has shown the potential for opening new vistas in science and medicine'

From Yalow's Nobel lecture (8 December 1977)

Yalow lived most of her life in New York City. She studied physics at Hunter College in the city and graduated in 1941. In the same year, she took up a teaching assistantship in physics at the University of Illinois at Urbana-Champaign, where she earned a PhD in nuclear physics in 1945. While teaching physics at Hunter College, she joined the Bronx Veterans Administration Hospital in 1947 to work on medical applications of radioisotopes. In 1968, she was appointed professor at the Mount Sinai Hospital.

In a long-term partnership with Solomon Berson, Yalow developed the technique of radioimmunoassay, a highly sensitive method to measure levels of antigens, chemicals in the bloodstream that trigger the release of antibodies. A known quantity of an antigen – for example, the hormone insulin – is made radioactive and mixed with a known amount of the corresponding antibody so that the two chemically bind together.

When this mixture is added to serum from a patient, a chemical competition begins as the radiolabelled and normal antigens compete for binding sites. When the bound and unbound antigens are later separated, it's possible to gauge the patient's antigen levels by measuring the bound sample's radioactivity levels with a gamma-ray detector (low levels of patient antigen will mean higher radioactivity because the radiolabelled antigen 'won' the binding competition). For this work, Yalow won a half share of the 1977 Nobel prize for physiology or medicine.

MARIA GOEPPERT MAYER
28 June 1906 – 20 February 1972

German–American physicist who developed the shell model of the atomic nucleus

'The shell model, although proposed by theoreticians, really corresponds to the experimentalist's approach. It was born from a thorough study of the experimental data, plotting them in different ways and looking for interconnections ... One of the main nuclear features which led to the development of the shell structure is the existence of what are usually called the magic numbers ... What makes a number magic is that a configuration of a magic number of neutrons, or of protons, is unusually stable whatever the associated number of the other nucleons.'

From Goeppert Mayer's Nobel lecture (12 December 1963)

From 1924, Maria Goeppert studied maths and physics at the University of Göttingen, where she was mentored by **Born** and developed a strong interest in the relatively new field of quantum mechanics. She earned her doctorate at Göttingen in 1930. In the same year, she married American chemist Joseph Edward Mayer and they moved to US where Goeppert Mayer's husband became professor at Johns Hopkins University in Baltimore, Maryland, then later at Columbia University and the University of Chicago.

Goeppert Mayer continued private research on chemical physics and worked in an unofficial or voluntary capacity at the universities where her husband was professor. 'This was the time of the depression, and no university would think of employing the wife of a professor. But I kept working, just for the fun of doing physics,' she later recalled. When the Argonne National Laboratory was founded in 1946, she took a part-time job there as a senior physicist and worked with **Urey** on techniques for separating uranium isotopes. She also collaborated with **Teller** and **Fermi**. In 1960, she moved to the University of California, San Diego.

Goeppert Mayer is best known for developing the shell model of the atomic nucleus. The shell model for the nucleus is analogous to the shell model of the atom. It aimed to explain the high stability of nuclei that have certain 'magic numbers' of either protons or neutrons inside them, specifically the numbers 2, 8, 20, 50, 82 and 126. Goeppert Mayer was able to explain their high stability by describing the nucleus as a series of closed shells in which pairs of neutrons and protons couple together due to an effect called spin orbit coupling. The magic numbers represent nuclei in which certain key shells are complete.

Goeppert Mayer summarized her model with this analogy: 'Think of a room full of waltzers. Suppose they go round the room in circles, each circle enclosed within another. Then imagine that in each circle, you can fit twice as many dancers by having one pair go anti-clockwise and another pair go counterclockwise. Then add one more variation; all the dancers are spinning twirling round and round like tops as they circle the room, each pair both twirling and circling. But only some of those that go counter-clockwise are twirling counterclockwise. The others are twirling clockwise while circling counterclockwise. The same is true of those that are dancing around clockwise: some twirl clockwise, others twirl counterclockwise.'

Independently, a team of scientists in Germany led by Hans Jensen developed the same idea. Goeppert Mayer contacted them to seek their collaboration and in 1950, she and Jensen jointly published a book on the subject, *Elementary Theory of Nuclear Shell Structure*. For her work on the shell model, Goeppert Mayer and Jensen shared half of the 1963 Nobel prize for physics. She was only the second woman after **Marie Curie** to win the physics prize, and the first woman to win it for theoretical work. The other half of the prize was awarded to the Hungarian–American physicist Eugene Paul Wigner for his contributions to nuclear theory and work on elementary particles.

In other research, Goeppert Mayer developed mathematical descriptions of optical opacity while working for Teller. Her work in this field contributed to the design of the first hydrogen (nuclear fusion) bomb, which was detonated in November 1952. After she died, the American Physical Society established an annual award in her name to recognize outstanding achievement by a young woman physicist and fund guest lectures about their research at four major institutions.

Italian neurologist who isolated nerve growth factor

'It was very good when I realized the importance of the
discovery of the nerve growth factor, but all together the
very end of my life gives me the possibility of work, not only
scientifically but socially, what I wanted when I was 20 years
old. I did not believe I would be a scientist. I wanted to go to
Africa to fight leprosy. This was my idea of life. At the very end
I work scientifically, but I work also to help women in the end.
So I do believe that the best period of my life is the present.'

From an interview with Levi-Montalcini aged 99
(*Annual Review of Physiology*, 2 September 2008)

Born to a Jewish family in Turin, Levi-Montalcini studied at the
medical school there, graduating with a degree in medicine and surgery
in 1936. Following Mussolini's introduction of laws barring Jews from
holding academic careers, she carried out research at a home laboratory
during the Second World War, and late in the war worked as a doctor
and nurse treating refugees. After that she was invited to Washington
University in St Louis, Missouri, where she later became professor until
her retirement in 1977.

Working in St Louis with American biochemist Stanley Cohen in the
1950s, Levi-Montalcini isolated nerve growth factor (NGF) after observing
extremely fast growth of nerve cells in certain cancerous tissues. NGF later
turned out to be a small protein vital for the growth, maintenance and
survival of certain nerve cells.

For the discovery of NGF, Levi-Montalcini and Cohen shared the 1986
Nobel prize for physiology or medicine. She has also been a committed
campaigner for the cause of alleviating disease, poverty and malnutrition.

FREDERICK SANGER
13 August 1918 –

English biochemist who determined the structure of insulin and developed techniques to sequence DNA

'We have been doing this not primarily to achieve riches or even honour, but rather because we were interested in the work, enjoyed doing it and felt very strongly that it was worthwhile. Scientific research is one of the most exciting and rewarding of occupations. It is like a voyage of discovery into unknown lands, seeking not for new territory but for new knowledge. It should appeal to those with a good sense of adventure.'

From Sanger's Nobel banquet speech (10 December 1980)

Sanger was born in the village of Rendcomb in Gloucestershire. He studied at Cambridge University, graduating with a first degree in 1939 and a PhD in biochemistry in 1943, having remained undrafted during the Second World War as a conscientious objector. From 1951 he was on the staff of the Medical Research Council in Cambridge.

In work at Cambridge beginning in 1943, Sanger developed new methods for sequencing the amino acids in proteins, and used them to work out the complete sequence for the hormone insulin. He went on to develop methods for sequencing RNA, a molecular cousin of DNA, and by 1967, his team had determined the nucleotide sequence of a small RNA molecule from *Escherichia coli* bacteria. Sanger then turned to sequencing DNA itself, and developed a fast technique for this that was eventually used to sequence the entire human genome.

For his work on insulin, Sanger won the 1958 Nobel prize for chemistry, and for his work on nucleic acids, he shared the same prize in 1980 with American scientists Walter Gilbert and Paul Berg. He remains the only person to have won a chemistry Nobel twice.

CHARLES HARD TOWNES
28 July 1915 –

American physicist who developed masers and laser theory

'It seems about time now for masers and lasers to become
everyday tools of science, and for the exploratory work which
has demonstrated so many new possibilities to be increasingly
replaced by much more finished, more systematic and more
penetrating applications'

From Townes's Nobel lecture (11 December 1964)

At the age of just 19, Townes had graduated from Furman
University in his hometown of Greenville, South Carolina with a degree
in physics and modern languages. He then went on to earn a physics
PhD at the California Institute of Technology in 1939. He worked at Bell
Telephone Laboratories from 1933 to 1947, and during the Second World
War he designed radar bombing systems. He later held professorships
at Columbia University in New York, the Massachusetts Institute of
Technology and the University of California at Berkeley.

In 1951, Townes conceived the concept of the maser (Microwave
Amplification by Stimulated Emission of Radiation) – the microwave
equivalent of the laser. His team developed the first working maser by
1954 and showed theoretically how to design optical and infrared lasers.
Both technologies excite atoms or molecules to higher energy levels in
a cavity, then trigger them to emit photons that have exactly the same
wavelength and are 'coherent', meaning the waves all travel in lockstep
with their crests and troughs lined up.

Lasers have found a host of everyday applications, such as reading
information in DVD players and delivering focused energy for delicate
surgery. For this work, Townes shared the 1964 Nobel prize for physics
with Russian physicists Nikolay Basov and Alexander Prokhorov. He has
also branched out into research on astronomy, and his team was first
to measure the mass of the supermassive black hole in the heart of the
Milky Way.

ROSALIND ELSIE FRANKLIN
25 July 1920 – 16 April 1958

Franklin was a key player in the discovery of the structure of DNA, life's molecular blueprint

'As a scientist Miss Franklin was distinguished by extreme clarity and perfection in everything she undertook – her photographs are among the most beautiful X-ray photographs of any substance ever taken'

From an obituary of Rosalind Franklin (*Nature*, 19 July 1958)

Born into a wealthy Jewish family in London, Franklin did a degree and PhD in chemistry at Cambridge University. After the Second World War, she worked at a research laboratory in Paris, which brought her up to speed in the art of X-ray diffraction. When shining through crystals, X-rays create complicated interference patterns that can be photographed to reveal the intricate arrangements of the atoms inside.

In 1951, Franklin moved to King's College, London, where she was asked to carry out X-ray crystallography of DNA, the molecule that encodes the genetic instructions for life. Another King's researcher Maurice Wilkins had already started this work. Between 1951 and 1953 Franklin came very close to solving the DNA structure, but she was beaten to publication by **Crick** and **Watson**, who won the race partly because Wilkins showed Watson one of Franklin's X-ray diffraction photos of DNA without her permission. However, there is some controversy over the extent of her contribution and whether or not the scientific community marginalized her, possibly motivated by sexism and personality clashes.

Franklin also clarified the structures of DNA's molecular cousin RNA as well as some viruses. But she died at the age of 37 from ovarian cancer. Four years later, Crick, Watson and Wilkins won the Nobel prize for physiology or medicine for their work on DNA.

JONAS EDWARD SALK
28 October 1914 – 23 June 1995

American virologist who developed the first
effective polio vaccine

'There are three stages of truth. First is that it can't be true and that's
what they said. You couldn't immunize against polio with a killed-
virus vaccine. Second phase: they say, "Well, if it's true, it's not very
important". And the third stage is, "Well, we've known it all along".
What you are describing is the process that you have to go through
when you come up with an idea that has not yet been tried or tested.'

From an interview with Salk at the Academy of Achievement

Salk studied medicine at New York
University, where he graduated MD in 1939. After
holding various research and teaching posts, in 1947
he joined the medical school at the University of
Pittsburgh in Pennsylvania. In 1963, he became the
founding director of the Salk Institute for Biological
Studies in San Diego, California.

After early research on flu vaccines, Salk turned
his attention to poliomyelitis, a highly feared disease to which children
were especially susceptible. The year 1952 saw the worst epidemic in the
US, with around 58,000 cases reported. More than 3,000 died of the
disease while more than 21,000 were left with mild to severe paralysis.
Salk confirmed that there were three types of virus that caused the disease
and he began experiments to try to kill the virus yet retain its ability to
provoke an immune response.

By 1952, Salk developed a vaccine that used poliovirus inactivated by
formalin, which when injected, conferred immunity against the three
types of polio virus and prevented the disease progressing. Years passed as
clinical trials of the vaccine were carried out, but by 1955, the vaccine was
shown to be safe and effective, leading to mass immunization campaigns.
In 1961, fewer than 200 polio cases were recorded in the US.

In the 1960s, the vaccine was superseded by a live-virus oral vaccine
developed by American medical researcher Albert Sabin. In his later years,
Salk worked on possible vaccines against HIV.

JACQUES-YVES COUSTEAU
(COMMONLY KNOWN AS JACQUES COUSTEAU)
11 JUNE 1910 – 25 JUNE 1997

French oceanographer, ecologist and filmmaker

'From birth, man carries the weight of gravity on his
shoulders. He is bolted to earth. But man has only to sink
beneath the surface and he is free. Buoyed by water, he can fly
in any direction – up, down, sideways – by merely flipping his
hand. Under water, man becomes an archangel.'

Cousteau quoted by *Time* magazine (28 March 1960)

Cousteau was born in Saint-André-de-Cubzac, Gironde, in southwest France. In 1930,
he entered the Brest Naval Academy and graduated
as a gunnery officer. He worked with the French
Resistance during the Second World War, for which
he was made a Commander of the Legion of Honour
and awarded the Croix de Guerre.

During the 1940s, Cousteau invented the
aqua-lung, a forerunner of modern scuba diving technology. For the
first time, divers could spend an extended period under water breathing
compressed air, without a heavy suit or lifeline. From 1950, he was
commander of the oceanographic research ship *Calypso*, from which he
pioneered underwater filming.

In his first book *The Silent World* (1953), Cousteau correctly proposed
after watching porpoise behaviour that they use echolocation to navigate.
His film with the same title (co-produced with Louis Malle) won the
prestigious Palme d'Or at the Cannes Film Festival in 1956. In 1957, he
became director of the Oceanographical Museum in Monaco.

Cousteau also discovered important shipwrecks and over his lifetime
made more than 120 television documentaries, published more than
50 books, and became a vigorous environmental campaigner. He also
contributed to the development of deep-sea submersibles. In 1985, he
received the US Presidential Medal of Freedom from Ronald Reagan.

FRANCIS HARRY COMPTON CRICK
8 June 1916 – 28 July 2004

British biologist and biophysicist who co-discovered the structure of DNA

'If for example I had some idea, which as it turned out would say be quite wrong, was going off at a tangent, Watson would tell me in no uncertain terms this was nonsense, and vice versa. If he would have some idea I didn't like, and I would say so, and this would shake his thinking about it and draw him back again. And in fact it's one of the requirements for collaboration of this sort, is you must be perfectly candid, one might almost say rude, to the person you're working with. It's useless working with somebody who is either much too junior than yourself or much too senior because then politeness creeps in. And this is the end of all real collaboration in science.'

From a BBC interview with Crick (*The Prizewinners*, 11 December 1962)

Born in Northampton, Crick studied physics at University College London, where he graduated in 1937. His PhD was interrupted in 1939 by the outbreak of the Second World War, during which he researched magnetic and acoustic mines for the British Admiralty. He left the Admiralty in 1947 to study biology at Cambridge, where he worked at the Strangeways Research Laboratory.

Two years later, he joined the Medical Research Unit at Cambridge's Cavendish Laboratory, where he worked with Austrian-born molecular biologist Max Perutz on determining the structure of proteins using X-ray crystallography. He earned his PhD in 1954 and remained at Cambridge until 1976.

When **Watson** joined the Cavendish Lab in 1951, the pair quickly became friends and began a project to tease out the structure of the DNA molecule, using Crick's expertise in X-ray diffraction and Watson's knowledge of bacterial genetics. Combining evidence from biochemistry and X-ray diffraction images created by **Rosalind Franklin** and Maurice Wilkins, as well as physical clues from molecular models, they showed that the 3D structure of DNA is a double helix. The landmark discovery was published in April 1953 in the journal *Nature* (they decided the order of the paper's authors, Watson and Crick, by flipping a coin).

Soon after, they published another paper suggesting how DNA replicates. Crick then attempted to find out how the genetic code works. In 1958, he proposed 'the sequence hypothesis' (that a DNA sequence codes for the amino acid sequence of a protein) and the 'central dogma' (that information goes from DNA to RNA to protein, but not back again). In 1957, Crick began work with **Brenner** to find out exactly how the sequence of DNA bases specifies the amino acid sequence in proteins. In 1961, they reported that translation must involve a three-nucleotide code.

Crick shared the 1962 Nobel prize for physiology or medicine with Watson and Wilkins for their work on DNA. In 1976, he became professor at the Salk Institute for Biological Studies in San Diego, California, and began studies of the brain and consciousness, hoping to spur progress by promoting collaboration between seemingly isolated experts from the many relevant fields.

He also considered many other topics, including panspermia, the idea that life did not originate on Earth, but was delivered to our planet by comets or asteroids. He even speculated that intelligent extraterrestrial civilizations brought life to Earth. He also wrote several popular books, including *Of Molecules and Men* (1967), which discusses the relationship between science and religion, and *The Astonishing Hypothesis* (1994), which describes his views on consciousness.

American biologist who co-discovered the structure of DNA

'This evening is certainly the second most wonderful moment
in my life. The first was our discovery of the structure of DNA.
At that time we knew that a new world had been opened and
that an old world which seemed rather mystical was gone ...
Good science as a way of life is sometimes difficult. It often is
hard to have confidence that you really know where the future
lies. We must thus believe strongly in our ideas, often to point
where they may seem tiresome and bothersome and even
arrogant to our colleagues. I knew many people, at least when
I was young, who thought I was quite unbearable.'

From Watson's Nobel banquet speech (10 December 1962)

Watson was born in Chicago, Illinois,
and studied zoology at the University of Chicago,
where he earned his first degree in 1947. He earned
his PhD from Indiana University in Bloomington
in 1950 after studying bacteriophage (viruses that
infect bacteria) and he continued this research in
postdoctoral work in Copenhagen.

In 1951, Watson took a research post at the
University of Cambridge's Cavendish Laboratory, where he returned to
after a spell at the California Institute of Technology from 1953 to 1955.
In 1956, he moved to Harvard University, where he was appointed full
professor in 1961, and from 1968, he was director of Cold Spring Harbor
Laboratory on Long Island, New York. In the early 1990s, he also headed
the Human Genome Project at the National Institutes of Health. He played
a key role in obtaining funding for the project and vigorously encouraged
cooperation between governments and leading geneticists.

In 1951, Watson met physicist and molecular biologist Maurice
Wilkins who was working on DNA at King's College, London. He
showed Watson the X-ray diffraction pattern of crystalline DNA. This
intrigued him enough to make him change his focus to the structural

chemistry of nucleic acids and proteins, which he began to study at Cambridge in collaboration with **Crick**.

In April 1953, Watson and Crick published a report proposing that the molecular structure of DNA is two cross-linked helical chains. They came to this conclusion after studying the X-ray diffraction patterns of DNA recorded by Wilkins and **Rosalind Franklin**, and taking into consideration rules developed by the Austrian–American biochemist Erwin Chargaff. In 1950, Chargaff had shown that the amounts of the organic bases adenine and thymine in DNA were roughly the same, as were the amounts of cytosine and guanine.

This strongly hinted that these bases pair up together, although Chargaff did not explicitly state this connection himself. But he met Watson and Crick at Cambridge in 1952 and explained his findings to them. Watson and Crick went on to show that the organic bases linked helical strands together with hydrogen bonds. After publishing the structure of DNA, they soon published another paper suggesting how DNA replicates, effectively by 'unzipping' the two strands, which then act as templates to form new DNA.

By the late 1950s, their proposed structure was experimentally confirmed. Further work by Crick and others showed that the genetic code is based on non-overlapping triplets of bases, called codons, allowing **Nirenberg** and others to finally decipher the genetic code, heralding the birth of modern molecular biology.

For their work on the structure of DNA, Watson, Crick and Wilkins shared the 1962 Nobel prize for physiology or medicine. Watson has written many books, including the bestseller *The Double Helix* (1968), which describes how the structure of DNA was unravelled with details of the personality clashes and difficulties along the way.

Watson has often expressed controversial views, arguing that if a gene that confers homosexuality or heterosexuality can be found, parents should have the right to choose their child's sexual orientation. In the *San Francisco Chronicle*, he was quoted as saying: 'Whenever you interview fat people, you feel bad, because you know you're not going to hire them.'

He was forced to retire from his post at Cold Spring Harbor Laboratory after an article in *The Times* reported his view that on average, people of African descent are less intelligent than people of European descent. He later released a backtracking statement in which he apologized unreservedly for the remarks, adding that they had no scientific basis.

STANLEY LLOYD MILLER
7 MARCH 1930 – 20 MAY 2007

American chemist and biologist who carried out a famous experiment to shed light on the origins of life

'Urey gave a lecture in October of 1951 when I first arrived at Chicago and suggested that someone do these experiments. So I went to him and said, "I'd like to do those experiments". The first thing he tried to do was talk me out of it ... He said the problem was that it was really a very risky experiment and probably wouldn't work, and he was responsible that I get a degree in three years or so. So we agreed to give it six months or a year ... As it turned out I got some results in a matter of weeks.'

Miller describes how his famous experiments on the origins of life came about
(from a 1996 interview)

Born in Oakland, California, Miller earned his first degree in 1951 from the University of California at Berkeley and a PhD in chemistry from the University of Chicago in 1954. After a brief spell at the California Institute of Technology in Pasadena, he joined Columbia University in New York before returning to the University of California, where he held a series of professorships.

In the 1950s, **Urey** suggested that the early atmosphere of the Earth was rich in ammonia, methane, and hydrogen – similar to the atmospheres of Jupiter and Saturn. Working in Urey's Chicago lab, Miller showed that when he zapped a similar mixture and water with an electrical discharge, the compounds can react to create the amino acids essential for living organisms. This could have played an important role in life's origins, but the composition of the Earth's early atmosphere remains speculative.

CLAUDE LÉVI-STRAUSS
28 November 1908 – 30 October 2009

*French anthropologist and philosopher who laid the
foundations of modern anthropology*

'I hate travelling and explorers ... Adventure has no place in the
anthropologist's profession; it is merely one of those unavoidable
drawbacks, which detract from his effective work through the
incidental loss of weeks or months; there are hours of inaction
when the informant is not available; periods of hunger, exhaustion,
sickness perhaps; and always the thousand and one dreary tasks
which eat away the days to no purpose and reduce dangerous living
in the heart of the virgin forest to an imitation of military service.'

From Lévi-Strauss's *Tristes Tropiques* (1955)

Lévi-Strauss was born in Brussels
and grew up in Paris, where he studied law and
philosophy at the Sorbonne, graduating in 1931.
After teaching in a secondary school, he joined a
French Cultural Mission team sent to Brazil, where
he taught at the University of São Paulo. During this
period he also travelled to remote regions of Brazil to
study the social structures of native tribes.

After serving in the French army during the Second World War and
holding several posts in France and the US, Lévi-Strauss was appointed
professor of social anthropology at the Collège de France in 1959. He
introduced and popularized structural anthropology, the notion that all
people and cultures think about the world in terms of binary opposites,
such as life and death, up and down, and animal versus human. He
argued that these opposing ideas create both conflict and also resolution
in cultural practices such as marriage and ritual.

In his work, Lévi-Strauss often used analogies from linguistics,
arguing that just as people are not naturally aware of the grammar of their
language, they are also not conscious of their day-to-day social structures.

HANS HUGO BRUNO SELYE
26 January 1907 – 16 October 1982

Austro-Hungarian endocrinologist who studied biological responses to stress

'I tried to demonstrate that stress is not a vague concept, somehow related to the decline in the influence of traditional codes of behaviour, dissatisfaction with the world, or the rising cost of living, but rather that it is clearly a definable biological and medical phenomenon whose mechanisms can be objectively identified and with which we can cope much better once we know how to handle it'

From Selye's treatise *The Nature of Stress* (1982)

Selye studied medicine and chemistry in Prague before moving to the US in 1931, where he studied at the Johns Hopkins University on a Rockefeller Foundation scholarship. He later went to McGill University in Montreal, where he began researching stress responses in 1936. In 1945, he moved to the University of Montreal as professor and director of the Institute of Experimental Medicine and Surgery.

In work on stress, a term that Selye coined, he emphasized that it is not a vague impression somehow related to discontentment with some aspect of life. Instead, he showed that it is a clearly definable biological and medical phenomenon whose mechanisms can be objectively identified. He identified key stages in the response, starting with alarm, when the body produces adrenaline to enable a 'fight-or-flight' response. A second 'resistance' stage is followed by 'exhaustion', when all of the body's resources are depleted, which can result in disorders including cardio-vascular disease.

Selye's work dramatically changed attitudes to disease and its causes, and he was nominated for a Nobel prize several times. His books included *From Dream to Discovery: On Being a Scientist* (1964) and *Stress without Distress* (1974).

WILLIAM ALFRED FOWLER
9 August 1911 – 14 March 1995

*American physicist who clarified how nuclear reactions
in stars create chemical elements*

'All of the heavy elements from carbon to uranium have been
synthesized in stars. Let me remind you that your bodies
consist for the most part of these heavy elements. Apart from
hydrogen you are 65 per cent oxygen and 18 per cent carbon
with smaller percentages of nitrogen, sodium, magnesium,
phosphorus, sulphur, chlorine, potassium, and traces of still
heavier elements. Thus it is possible to say that you and your
neighbour and I, each one of us and all of us, are truly and
literally a little bit of stardust.'

From Fowler's Nobel lecture (8 December 1983)

Fowler was born in Pittsburgh, Pennsylvania. He studied
ceramic engineering at Ohio State University in Columbus, where he
also took classes in physics and maths then switched to engineering
physics. After that he did a doctoral thesis on radioactive elements at the
California Institute of Technology in Pasadena, where he later became
professor in 1946.

Working with British astronomers Geoffrey and Margaret Burbidge
as well as **Hoyle**, Fowler published a landmark paper in 1957 showing
that all of the heavy elements in the Universe from carbon to uranium
can be produced by nuclear processes in stars starting with the hydrogen
and helium produced in the Big Bang. Previously, **Gamow** had proposed
that all nuclei heavier than helium could have been built up in the Big
Bang itself.

Fowler later did experimental work to quantify nuclear reaction rates,
an essential step in confirming stellar nucleosynthesis quantitatively. For
his work on the synthesis of elements, he shared the 1983 Nobel prize for
physics with **Chandrasekhar.**

PHILIP WARREN ANDERSON
13 December 1923 –

American physicist who developed theories of magnetic and disordered materials

'String theory is the first science in hundreds of years to
be pursued in pre-Baconian fashion, without any adequate
experimental guidance. It proposes that Nature is the way we
would like it to be rather than the way we see it to be; and it is
improbable that Nature thinks the same way we do.'

Anderson explains his objections to string theory in *The New York Times*
(4 January 2005)

Born in Indianapolis, Indiana, Anderson grew up in Urbana,
Illinois. He studied maths at Harvard University and during the wartime
years of 1940 to 1943, he was encouraged to focus on electronics; following
that he built antennas at the Naval Research Laboratory in Washington
DC. After the war he returned to Harvard to do his PhD, which he earned
in 1949.

Working at Bell Telephone Laboratories in New Jersey, Anderson
studied disordered materials (as opposed to ideal crystals with atoms
arranged at repeating, fixed distances) and showed that electrons in them
can become trapped in a process now called 'Anderson localization'. This
understanding led to extensive use of non-crystalline semiconductors
in technologies such as solar cells and transistors. In the 1960s, he
also clarified the nature of possible superfluid states of helium-3, later
discovered in 1972.

For his work, Anderson won a third share of the 1977 Nobel prize
for physics. He has been a vocal critic of string theory, which attempts
to reconcile quantum mechanics with general relativity by modelling
fundamental particles like electrons and quarks as mini vibrating strings.
He mourns the 'incredible amount of effort' expended on the theory,
despite the fact that it makes no testable predictions. Anderson retired
from Bell Labs in 1984 and became professor of physics at Princeton
University in New Jersey.

ALLAN REX SANDAGE
18 JUNE 1926 – 13 NOVEMBER 2010

American astronomer who determined the first reasonably accurate age of the Universe

'It was my science that drove me to the conclusion that the world is much more complicated than can be explained by science. It is only through the supernatural that I can understand the mystery of existence.'

Sandage quoted in 'Science Finds God' (article by Sharon Begley in *Newsweek*, 20 July 1998)

Sandage was born in Iowa City. He graduated from the University of Illinois in 1948 and earned a PhD from the California Institute of Technology in 1953 after serving as an observing assistant to **Hubble** at the Mount Wilson and Palomar observatories. From 1952, he was on the staff of the Carnegie Observatories based in Pasadena.

When Hubble died in 1953, Sandage became responsible for developing the Carnegie cosmology programme that aimed to measure distance scales in the Universe. He showed that Hubble's distances to galaxies had been hugely underestimated and in 1958 published the first reasonably accurate measurement of the so-called 'Hubble constant', from which astronomers can estimate the age of the Universe. Sandage's work suggested an age of 15 billion years, not far off today's established age of 13.7 billion years.

In work on stellar evolution, Sandage developed methods to estimate the ages of stars. He also made the first identification at optical wavelengths of quasars, now thought to be highly energetic galaxies powered by supermassive black holes. A prolific researcher, he published more than 500 papers during his career. Late in life, he became a committed Christian and wrote essays on religion and science.

SYDNEY BRENNER

13 JANUARY 1927 –

South African biologist who clarified the genetic code and details of animal development

'Without doubt the fourth winner of the Nobel prize this
year is [the nematode worm] *Caenorhabditis elegans*; it
deserves all of the honour but, of course, it will not be
able to share the monetary award'

From Brenner's Nobel lecture (8 December 2002)

Born to Jewish immigrants in the small town of Germiston,
South Africa, Brenner showed early scientific talent and aged just 15
received a bursary from his town council to study science and medicine
at the University of the Witwatersrand in Johannesburg. He moved to the
University of Oxford in the UK where he earned his doctorate in 1954.
From 1956 he worked at Cambridge University, becoming director of the
Medical Research Council's Laboratory for Molecular Biology in 1979.

Brenner describes the discovery of the structure of DNA by **Crick** and
Watson as 'the watershed in my scientific life' adding: 'I realized that it
was the key to understanding all the problems in biology we had found
intractable'. In the late 1950s and the 1960s, he made key contributions
to the unravelling of the genetic code, confirming the triplet nature of the
DNA base coding for amino acids and showing that the triplets (codons)
don't overlap (so a linear sequence of six nucleotides can only encode two
amino acids, not four).

He also established the tiny soil roundworm *Caenorhabditis elegans* as
a model organism for investigating animal development, chosen because
it's simple, easy to breed and convenient for genetic analysis. His work
on *C. elegans* clarified the genetic processes involved in aging, nerve cell
function and controlled cell death. For this work, Brenner won a third
share of the 2002 Nobel prize for physiology or medicine.

JANE MORRIS GOODALL
3 April 1934 –

British primatologist best known for studying social interactions of wild chimpanzees

'I, with my work at Gombe, was trying to understand a little better the nature of chimpanzee aggression. My question was: how far along our human path, which has led to hatred and evil and full-scale war, have chimpanzees travelled?'

From Goodall's *Reason for Hope* (2004 edition)

Goodall had a childhood passion for animals and in 1957 travelled to Kenya, where she was employed by **Leakey** as a secretary. The following year Leakey arranged for her to study primate anatomy and behaviour in London and he raised funds for her to begin a study of wild chimps at Gombe Stream National Park in Tanzania. She also enrolled at Cambridge University for a PhD in ethology, which she earned in 1965.

With her colleagues at Gombe, Goodall carried out the longest continuous field study of animals in their natural habitat. Her work radically changed the prevailing picture of the social interactions between chimps, revealing that they have unique, individual personalities and demonstrate a wide range of human-like emotions and behaviours, such as kissing and tickling to maintain supportive, affectionate bonds. She also recorded the first known examples of chimps modifying objects like plant stems as tools to collect termites, and rocks as missiles to fend off predators.

Goodall also showed that chimps eat meat, strategically hunting smaller primates such as colobus monkeys, with the chimp troop sharing the carcasses. She founded the Jane Goodall Institute in 1977 to promote the study and protection of chimps, and to empower people to improve the environment for all living things. While estimates suggest there were around a million wild chimpanzees at the turn of the 20th century, today there are fewer than 200,000.

HARRY HAMMOND HESS
24 May 1906 – 25 August 1969

American geologist who developed early theories of plate tectonics

'In order not to travel any further into the realm of fantasy than is absolutely necessary I shall hold as closely as possibly to a uniformitarian approach; even so, at least one great catastrophe will be required early in the Earth's history'

Hess quoted in 'History of Ocean Basins'
(*Petrologic Studies: A Volume to Honour F. Buddington*, 1962)

Hess was educated at Yale University in Connecticut and Princeton University, New Jersey, where he earned a PhD in 1932. After a brief spell teaching at Rutgers University in New Jersey, he spent a year as a research associate at the Geophysical Laboratory of Washington DC and joined the faculty of Princeton University in 1934. For the rest of his career, Hess remained at Princeton where he was chair of the geology department from 1950 to 1966.

A strong influence on Hess's career was service in the US Navy during the Second World War, when he was captain of the transport ship USS *Cape Johnson*. The ship was equipped with sonar equipment that allowed him to measure ocean floor depths across the North Pacific. The observations revealed flat-topped volcanoes on the ocean floor, which he named guyots. He stayed in the navy after the war and was promoted to the rank of rear admiral.

In a 1960 report for the US Office of Naval Research, Hess suggested correctly that the Earth's crust moves outward from volcanically active ridges on the ocean floor, a process now called seafloor spreading. The suggestion filled in gaps in **Wegener**'s earlier theory of continental drift by introducing a plausible mechanism for it, and this played a key role in the continental drift theory gaining widespread acceptance.

FRANK DONALD DRAKE
28 May 1930 –

*American astrophysicist who pioneered the search for
extraterrestrial intelligence (SETI)*

'Back in 1960 it was taboo to think about extraterrestrial life; it
was something done by bad scientists. However, we were fearless.
We did not feel we should be embarrassed in any way.'

From an interview with Drake in *New Scientist* magazine (13 January 2010)

Drake was born in Chicago and won a navy
scholarship to study engineering physics at Cornell
University, where he graduated in 1952. After that
he earned a PhD in radio astronomy from Harvard
in 1958. He then worked at the National Radio
Astronomy Observatory in West Virginia before
holding several posts at Cornell University between
1964 and 1984. Since then, Drake has worked at
the SETI Institute in Mountain View, California, and the University of
California, Santa Cruz.

In 1960, Drake began the first modern SETI project, named 'Ozma'.
Using a large radio telescope, he monitored the nearby Sun-like stars Tau
Ceti and Epsilon Eridani to look for radio broadcasts from intelligent civilizations on any planets that might be orbiting the stars. With **Sagan**, he
also designed the 'Arecibo message', a radio transmission beamed into
outer space from the giant Arecibo radio telescope in Puerto Rico on
16 November 1974. The message encoded information about our solar
system and life on Earth.

But Drake is best known for formulating the Drake equation, which
expresses the number of civilizations in our galaxy with which communication might be possible as the product of seven factors, including the
fraction of stars that have planets and the fraction of civilizations capable
of transmitting detectable signals. He is still involved in ongoing SETI
projects in California.

SHELDON LEE GLASHOW
5 December 1932 –

American physicist who unified the electromagnetic and weak forces in a single theory

'When I began doing theoretical physics, the study
of elementary particles was like a patchwork quilt.
Electrodynamics, weak interactions, and strong interactions
were clearly separate disciplines, separately taught and
separately studied. There was no coherent theory that
described them all ... The theory we now have is an integral
work of art: the patchwork quilt has become a tapestry.'

From Glashow's Nobel banquet speech (8 December 1979)

Born to a Russian Jewish family in New York City, Glashow was
a childhood friend of **Weinberg** and studied physics at Cornell University
and Harvard, where he earned his PhD in 1959. He held posts at Stanford
University in California and the University of California at Berkeley before
in 1966 becoming professor at Harvard, where he is currently Higgins
Professor of Physics (Emeritus).

Glashow developed an early theory to describe two of nature's
fundamental forces, the electromagnetic and weak forces, within a
single framework. Weinberg and **Salam** successfully applied a unified
'electroweak' theory to interactions between leptons, the group of
fundamental particles (including the electron) that don't feel the strong
force. Glashow then extended the theory to encompass all the known
matter particles at the time, including particles such as neutrons and
protons that are built from quarks.

The key to the success of Glashow's theory was the addition of the
property of 'charm'; he added a fourth quark called charm to Gell-Mann's
scheme of three quarks (today there are six known quarks). The existence
of this quark was confirmed in 1974 when **Ting** and **Richter** discovered the
J/ψ particle, made of a charm quark and antiquark. For his work, Glashow
shared the 1979 Nobel prize for physics with Salam and Weinberg.

MARSHALL WARREN NIRENBERG
10 April 1927 – 15 January 2010

American biochemist who cracked the genetic code

'Most, perhaps all, forms of life on this planet use essentially the same genetic language, and the language is translated according to universal rules ... The code was frozen after organisms as complex as bacteria had evolved because major alterations in the code would affect the amino acid sequence of most proteins synthesized by the cell and probably would be lethal.'

From Nirenberg's Nobel lecture (12 December 1968)

Born in New York City, Nirenberg developed an early interest in biology and studied zoology at the University of Florida at Gainesville, where he earned a bachelor's degree in 1948 and a master's degree in 1952. He then did a PhD in biochemistry at the University of Michigan, Ann Arbor, graduating in 1957. For most of his career, he worked at the National Institutes of Health in Maryland.

In 1961, working with postdoctoral fellow Heinrich Matthaei, Nirenberg cracked the genetic code by using nucleic acids to translate specific amino acids. They showed that by adding an artificial form of RNA (the molecular cousin of DNA) to an extract from bacterial cells, they could make the extract produce a protein composed entirely of the amino acid phenylalanine. This proved for the first time that RNA controlled the production of specific types of protein and that the sequence of three uracil nucleotides (UUU) was the blueprint for phenylalanine production.

His team went on to show that AAA (three adenosines) was the 'code word' for the amino acid lysine, while CCC (three cytosines) coded for proline. They also discovered that by replacing one or two units of a triplet with other nucleotides, they could direct the production of other amino acids. By 1966, Nirenberg announced that he had deciphered the 64 RNA codons for all 20 amino acids that make up proteins. For this work, he won a share of the 1968 Nobel prize for physiology or medicine.

LEON MAX LEDERMAN
15 JULY 1922 –

American physicist who discovered two of nature's fundamental particles

'Energy, if you keep track it, it should balance. And in these reactions, it didn't balance. Something was missing. For a while, physicists jumped out of second-storey windows, they got very upset because they really loved conservation of energy, and it looked as if they were going to lose it as a principle.'

Lederman describes the clues that undetected particles (neutrinos) steal energy from experiments (from a 1992 interview by the Academy of Achievement)

Lederman studied physics at Columbia University in New York, where he earned his PhD in 1951. He remained at Columbia for 28 years, becoming full professor in 1958 and in 1979, he was appointed director of the Fermi National Accelerator Laboratory near Chicago, where he supervised the construction and management of the highest energy particle accelerator of its day.

By the 1960s, scientists knew that the fundamental particles in nature include the neutrino, a slippery neutral particle that's nearly impossible to detect because it flies straight through detectors without interacting. There were hints that the neutrinos might come in at least two 'flavours', electron and muon, and an experiment at Brookhaven National Laboratory by Lederman, Melvin Schwartz and Jack Steinberger confirmed this in 1962.

Lederman also led a team that discovered the 'bottom quark', one of the six quarks now known in the standard model of particle physics. For their work on neutrinos, Lederman, Schwartz and Steinberger shared the 1988 Nobel prize for physics. Lederman's lively sense of wit comes across in his popular science book *The God Particle: If the Universe is the answer, what is the question?* (1993).

MURRAY GELL-MANN
15 September 1929 –

American physicist who proposed the existence of
fundamental particles called quarks

'The beauty of the basic laws of natural science, as revealed in the
study of particles and of the cosmos, is allied to the litheness of a
merganser diving in a pure Swedish lake, or the grace of a dolphin
leaving shining trails at night in the Gulf of California'

From Gell-Mann's Nobel banquet speech (10 December 1969)

Born in New York City, Gell-Mann began
studying physics at Yale University when he was
only 15. He earned his PhD from the Massachusetts
Institute of Technology in 1951. Since 1955, he has
held posts at the California Institute of Technology,
where he is now Robert Andrews Millikan Professor
Emeritus. He is also one of the founders of the
Santa Fe Institute in New Mexico, set up in 1984 to
conduct theoretical research into complex adaptive systems (for instance,
the emergence of early life on Earth).

In particle physics, Gell-Mann developed the 'eightfold way' scheme,
in essence a kind of 'periodic table' of subatomic particles, around 100
of which had been observed in the debris of collisions between particles
at accelerators. Gell-Mann correctly predicted that many subatomic
particles, including the neutron and proton, are composed of building
blocks he named quarks, which curiously can't be knocked out of their
parent particles because as they separate, the forces between them
become stronger.

This quark model has since been confirmed by experiments. Gell-Mann
and others also developed quantum chromodynamics, the quantum field
theory of quarks and the 'gluons' that mediate the strong force between
them. For his work on elementary particles, Gell-Mann won the 1969
Nobel prize for physics.

WILLIAM DONALD HAMILTON
1 August 1936 – 7 March 2000

*British evolutionary biologist who introduced a
modern theory of kin selection*

'If you believe that we evolved out of animals – are animals
– and have the same kinds of drives, it doesn't mean that we
have to be selfish and inhumane. When you fully work out
the consequences of the rules of kinship and of reciprocation
... the outcome is in fact quite a moderate kind of behaviour,
avoiding evil and as good in holding the society together as are
the religious myths.'

From an interview with Hamilton at the University of Oxford
(*Human Ethology Bulletin*, 1996)

Hamilton developed an interest in natural history, particularly
insects, from a young age. Born in Cairo, Egypt, but raised in Kent, he was
educated at the University of Cambridge and UCL. He taught in London
and at the University of Michigan before becoming professor at Oxford
University from 1984 until his death.

His major contribution was to provide a firm theoretical footing for
some of the problems in modern evolutionary biology. His early work
demonstrated the genetic basis of some social behaviour. In 1964, he
introduced a theory of kin selection that accounted for the altruistic
behaviour seen in some animal communities – even a sterile individual
may promote the survival and successful breeding of a relative to increase
the chances that shared genes will be passed on to the next generation.

He also determined the conditions necessary for a gene for altruistic
self sacrifice to spread through a population, and he studied the evolution
of sexual reproduction, describing it in terms of an elaborate mechanism
for combating parasites in endlessly evolving ecosystems. His collected
papers were published from 1996 as *Narrow Roads of Gene Land, Evolution
of Sex* and *Last Words*.

PETER WARE HIGGS
29 May 1929 –

English theoretical physicist who explained the possible origin of the mass of fundamental particles

'If I'm wrong, I'll be rather sad. If it is not found, I no longer understand what I think I understand.'

Higgs anticipates the discovery of the 'Higgs boson'
(from an interview with *The Daily Telegraph*, 8 April 2008)

Higgs was born in Newcastle upon Tyne and studied physics at King's College in London where he earned his PhD in 1954. After various posts at the University of Edinburgh, as well as University College and Imperial College in London, he returned to Edinburgh University in 1960 and became professor there in 1980.

In his work on particle physics during the 1960s, Higgs developed the idea that particles must have originally been massless just after the Big Bang that created the Universe about 13.7 billion years ago. A tiny fraction of a second later, however, fundamental particles such as quarks (including the constituents of protons and neutrons) and leptons (which include the electron) must have acquired mass by interacting with a theoretical field now called the Higgs field.

Associated with the Higgs field there would be a 'Higgs boson', a hypothetical massive particle that scientists are currently attempting to detect in experiments using the Large Hadron Collider at CERN (the European centre for particle physics on the Swiss–French border). This elusive particle has been nicknamed the 'God particle', though not by Higgs himself. He has also been a political activist, opposing Israel's stance on Palestine and supporting both the Campaign for Nuclear Disarmament and Greenpeace until they changed their agendas to lobby against nuclear power and genetically modified organisms.

ROGER PENROSE
8 August 1931 –

English mathematical physicist renowned for his work on general relativity and cosmology

'Classical physics is not explained by quantum physics. It's rather a patchwork: quantum physics on one side and classical on the other. When you look at it in all its implications, you realize that it's a totally unsatisfactory and incomplete picture of what must be going on in the world.'

Interview with Penrose in the subscription newsletter *Science Watch* (August 1991)

In 1939, Penrose emigrated with his parents from Colchester, where he was born, to Ontario in Canada, where the family remained until the end of the Second World War. He studied maths at University College London and Cambridge University, where he received a PhD in 1957. He held posts at various universities before in 1973 becoming Rouse Ball Professor of Mathematics at the University of Oxford, where he remained until his retirement in 1998. He was knighted in 1994.

Working at Cambridge with **Hawking** on **Einstein**'s general theory of relativity, Penrose showed in 1965 that 'singularities' can form when the core of a very massive star collapses at the end of the star's life. This leaves behind a black hole, a point of such enormous density that nothing – not even light – can escape its strong gravity. They also showed that general relativity implies that space and time began with the Big Bang.

Penrose has sought a more complete picture of nature's laws by working on quantum gravity theories, which would unite quantum theory with general relativity. He also created the 'impossible' Penrose stairs, a two-dimensional staircase with four 90-degree turns as they ascend or descend, yet in a continuous loop (it would be impossible in 3D). His report on this, published in 1959, inspired the Dutch artist Maurits Escher's famous lithograph *Ascending and Descending*.

ROBERT WOODROW WILSON
10 January 1936 –

American astronomer who co-discovered the cosmic microwave background, the relic radiation of the Big Bang

'Cosmology is a science which has only a few observable facts to work with. The discovery of the cosmic microwave background radiation added one – the present radiation temperature of the Universe.'

From Wilson's Nobel prize lecture (1978)

Wilson was an enthusiastic electronics dabbler as a child, keen on fixing radio equipment and TVs. He studied physics at Rice University in Houston and then did a PhD in radio astronomy at the California Institute of Technology in Pasadena.

Wilson joined Bell Laboratories at Crawford Hill in New Jersey in 1963 and began work with **Penzias** to monitor radio emissions from the Milky Way. Together, they serendipitously discovered the cosmic microwave background, effectively the afterglow of the Big Bang that created the Universe. For this discovery, Wilson shared half of the 1978 Nobel prize for physics with Penzias.

The discovery of the microwave background has transformed cosmology from mere speculation into a testable, experimental science. In 1991, NASA's COBE (Cosmic Background Explorer) satellite revealed that the microwave background contains subtle 'ripples' corresponding to density variations in the early Universe (see **Smoot**). Measurements of the ripples have allowed astronomers to study the geometry of the Universe and show accurately that the Universe is 13.7 billion years old.

Through their work in radio astronomy, Wilson and Penzias also discovered several interstellar molecules including deuterated hydrogen cyanide. This has allowed astronomers to map the interstellar distribution of deuterium (heavy hydrogen), which has added weight to the Big Bang theory because the theory suggests the deuterium was manufactured exclusively in the Big Bang itself. They also uncovered surprisingly large amounts of carbon monoxide in dusty interstellar molecular clouds behind the Orion Nebula.

ARNO ALLAN PENZIAS
26 April 1933 –

American astronomer who co-discovered the cosmic microwave background, the relic radiation of the Big Bang

'If you don't want to be replaced by a machine,
don't try to act like one!'

From Penzias's Nobel prize autobiography

Born in Munich into a Jewish family, Penzias fled to England from Nazi Germany in 1939 then on to New York the following year. He became a US citizen in 1946. He studied physics at the City College of New York then became a research assistant at Columbia University before doing a PhD there building radio astronomy equipment, supervised by **Townes**.

From 1961, he worked at Bell Laboratories in New Jersey for 37 years. Working with **Robert Wilson**, Penzias discovered the cosmic microwave background, supporting the theory of the Big Bang. The theory held that the Universe was born in a hot fireball that gradually expanded and cooled. **Alpher** and others had predicted in 1948 that if this picture was correct, space must be filled with an 'afterglow' stretched to microwave frequencies by the expansion of the Universe over time.

At Bell Labs, Penzias and Wilson used a 6 m (20 ft) horn-shaped antenna in 1964 and 1965 to map signals from the Milky Way. They were surprised to find a persistent 'noise' of microwaves coming from every direction. They tried to rule out possible noise sources such as urban interference or military radiation sources, and even wondered if the noise was due to droppings from pigeons nesting in their equipment. Robert Dicke, an American physicist, identified it as the predicted cosmic microwave background.

Penzias and Wilson shared half of the 1978 Nobel prize for physics. Penzias published *Ideas and Information* in 1989, arguing that computers are a wonderful tool for human beings but a dreadful role model for what we know as intelligence, hence the quote above.

LYNN MARGULIS

5 March 1938 – 22 November 2011

American biologist who recognized that organisms engulfed each other to create complex cells

'All beings alive today are equally evolved. All have survived over three thousand million years of evolution from common bacterial ancestors. There are no "higher" beings, no "lower animals" … We *Homo sapiens sapiens* and our primate relations are not special, just recent: we are newcomers on the evolutionary stage. Human similarities to other life forms are far more striking than the differences. Our deep connections, over vast geological periods, should inspire awe, not repulsion.'

From Margulis's *Symbiotic Planet: a New Look at Evolution* (1998)

Margulis attended the University of Chicago from the age of 14 and married **Sagan** at 19. She earned a master's degree from the University of Wisconsin in 1960 and a PhD in 1963 from the University of California at Berkeley. From 1966 to 1986 she held a series of posts at Boston University and in 1988, she became professor at the University of Massachusetts in Amherst, where she remained for the rest of her career.

In 1966, Margulis wrote a theoretical paper backing the 'endosymbiotic' theory, suggesting certain organelles in cells – such as the mitochondria that act as cell power houses – originated as free-living bacteria that were engulfed by cells. Instead of being completely digested, the bacteria thrived inside the cells in a mutually beneficial relationship and reproduced as the cell divided. Through many such events, eukaryotic cells that make up complex plants and animals evolved over millions of years.

Although the idea was unpopular at the time, it is now widely accepted. Margulis also developed **Lovelock**'s Gaia concept, which proposes that interactions on the Earth's surface between living organisms, sediment, air and water have created a vast self-regulating system.

(SUSAN) JOCELYN BELL BURNELL
15 JULY 1943 –

Northern Irish astrophysicist who co-discovered the first known pulsar, one of the most exotic objects in the Universe

'The source was well outside the solar system but inside the galaxy – so were these pulsations man-made, but made by man from another civilization?'

From Bell Burnell's after-dinner speech at an astrophysics symposium (1976)

Bell (later Bell Burnell after marriage) grew up in Lurgan, Northern Ireland, and York before graduating in physics from Glasgow University in 1965. Working with **Hewish** at Cambridge University, she earned her PhD in 1968 and later she held research, lecturing posts and professorships at various universities.

In July 1967, using a radio detector array near Cambridge, she noticed a curiously regular signal every 1.3 seconds. It seemed artificial, yet it was coming from outer space, not the Earth. Could it be aliens trying to communicate with us? Tongue-in-cheek astronomers, her team temporarily named the source LGM-1, short for Little Green Men 1. 'We did not really believe that we had picked up signals from another civilization, but obviously the idea had crossed our minds and we had no proof that it was an entirely natural radio emission,' Bell Burnell recalled in a 1976 speech.

After several years, astronomers identified the radio source (now known as PSR B1919+21) as a rapidly rotating neutron star, or pulsar. Neutron stars are the superdense collapsed cores of massive stars sometimes left behind after a very massive star explodes. Typically about 15 km (9 miles) wide, they emit bright beams of radiation from their poles, and these beams can sweep across the Earth like lighthouse beams as the neutron star spins. That explains the regular radio blips.

ANTONY HEWISH
11 May 1924 –

English radio astronomer who co-discovered the first known pulsar

'The world of man lies midway in scale between the inner space
of atoms and particles, and the outer space of stars and galaxies –
the exploration of both these regions stretches our
imagination to its limits'

From Hewish's Nobel banquet speech (10 December 1974)

Hewish began a degree in physics at Cambridge University
in 1942, but didn't graduate until 1948. His undergraduate degree was
disrupted by war service at the Royal Aircraft Establishment, Farnborough,
and also at the Telecommunications Research Establishment, Malvern,
when he worked on airborne devices to counter radar surveillance.

His wartime experience with electronics and antennas encouraged
him into a career in radio astronomy, which he later pursued during a PhD
at Cambridge's Cavendish Laboratory. Hewish recognized that he could
study the interplanetary medium by observing distant, energetic galaxies
called quasars using a radio telescope array. As point-like radio sources,
the 'scintillation' of quasars (flickering due to interplanetary gas motions)
made them the radio equivalent of visible stars that twinkle due to motion
of the Earth's atmosphere.

So Hewish arranged to build a large array radio telescope at the Mullard
Radio Astronomy Observatory near Cambridge for this purpose. In 1967,
one of his students on the project noticed a curious, repeating radio signal
that turned out to come from the first known pulsar, an exotic collapsed
star rotating once every 1.3 s (see entry for **Jocelyn Bell Burnell** for more
details). Hewish remained at Cambridge as professor of radio astronomy
from 1971 to 1989. For discovering pulsars, he won a half share of the
1974 Nobel prize for physics.

STEVEN WEINBERG
3 MAY 1933 –

American physicist who theoretically unified nature's electromagnetic and weak forces

'In the seventh book of the *Republic*, Plato describes prisoners
who are chained in a cave and can see only shadows that
things outside cast on the cave wall ... We are in such a cave,
imprisoned by the limitations on the sorts of experiments we
can do ... We have not been able to get out of this cave, but
by looking long and hard at the shadows on the cave wall,
we can at least make out the shapes of symmetries, which
though broken, are exact principles governing all phenomena,
expressions of the beauty of the world outside.'

From Weinberg's Nobel lecture (8 December 1979)

Born to a first-generation Jewish family in New York City,
Weinberg studied physics at Cornell University, where he earned his first
degree in 1954. He completed a PhD in 1957 at Princeton University in
New Jersey. After posts at several universities, he held professorships at
the Massachusetts Institute of Technology from 1969, Harvard University
from 1973 and the University of Texas from 1982.

In a series of separate developments in the 1960s, Weinberg, **Glashow**
and **Salam** developed the electroweak theory that describes nature's weak
and electromagnetic forces in a common framework. Effectively, it achieved
a long-standing goal of explaining how two fundamental forces acted as a
single one in the hot and energetic young Universe. For this work, Weinberg,
Glashow and Salam shared the 1979 Nobel prize for physics.

Weinberg has also been a prominent public spokesman for science
and he has written many popular science books including *The First Three
Minutes* (1977), which explains the dramatic aftermath of the Big Bang that
created the Universe.

ABDUS SALAM
29 January 1926 – 21 November 1996

Pakistani physicist who developed a unified theory of nature's electromagnetic and weak forces

'Alfred Nobel stipulated that no distinction of race or colour will determine who received of his generosity. On this occasion, let me say this to those, whom God has given His Bounty. Let us strive to provide equal opportunities to all so that they can engage in the creation of physics and science for the benefit of all mankind.'

From Salam's Nobel banquet speech (10 December 1979)

 **Salam was born in Jhang, India** (now in Pakistan). He was educated at Punjab University and the University of Cambridge, where he earned a PhD in theoretical physics in 1952. After holding posts in Lahore and Cambridge, he was appointed professor at Imperial College London in 1957 and director of the International Centre for Theoretical Physics, Trieste, in 1964. From 1961 to 1974, he was also chief scientific adviser to the president of Pakistan.

During the 1960s, Salam, **Weinberg** and **Glashow** developed the electroweak theory that describes nature's weak and electromagnetic forces in a common framework. The theory clarified how two of nature's four fundamental forces acted as a single force when the Universe was very young and energetic.

The theory also predicted the existence of so-called 'neutral currents', interactions mediated by a particle called the Z boson. These interactions were experimentally confirmed in 1973 at CERN, the European centre for particle physics on the French–Swiss border. For their work on the electroweak theory, Salam, Weinberg and Glashow shared the 1979 Nobel prize for physics.

ROBERT GEOFFREY EDWARDS
27 September 1925 –

British physiologist who pioneered in vitro *fertilization*

'The whole world was opening up. It's a revolution in thinking –
that man is controlling his own conception.'

Edwards recalls the birth of the first test-tube baby
(interview with BBC, 24 July 2003)

Edwards studied agricultural sciences
and zoology at the University College of North Wales
in Bangor before investigating the developmental
biology of the mouse at Edinburgh University, where
he earned a PhD in 1955. In 1963, he joined the
University of Cambridge, where he spent most of the
rest of his career.

As early as the 1950s, Edwards became
convinced that *in vitro* fertilization (IVF) could be useful as a treatment for
infertility. For many years, he studied the important principles for human
fertilization, clarifying how human eggs mature and the time at which
they're susceptible to fertilization by sperm. He succeeded in fertilizing a
human egg with sperm in a cell culture dish in 1969. Funding problems
as well as religious and ethical objections threatened to halt the project,
but Edwards's work with surgeon Patrick Steptoe finally resulted in a
healthy IVF pregnancy, leading to the birth of the first 'test-tube baby',
Louise Brown, on 25 July 1978.

Edwards and Steptoe went on to establish the first IVF clinic at Bourn
Hall in Cambridge in 1980. By 1988, more than 1,000 babies had been
conceived at the clinic and today, more than four million children around
the world have been born through IVF. For this work, Edwards won the
2010 Nobel prize for physiology or medicine.

EDWARD OSBORNE WILSON
10 JUNE 1929 –

American biologist considered the father of sociobiology

'Few will doubt that humankind has created a planet-sized
problem for itself. No one wished it so, but we are the first
species to become a geophysical force, altering Earth's climate,
a role previously reserved for tectonics, Sun flares and glacial
cycles ... Through overpopulation we have put ourselves in
danger of running out of food and water. So a very Faustian
choice is upon us: whether to accept our corrosive and
risky behaviour as the unavoidable price of population and
economic growth, or to take stock of ourselves and search for
a new environmental ethic.'

From E.O. Wilson's *Consilience: The Unity of Knowledge* (1998)

Wilson was born in Birmingham, Alabama, and developed
an interest in insects in childhood. He studied at the University of
Alabama and at Harvard, where he earned a PhD in entomology in 1955.
The following year, he joined the Harvard faculty and he remained there
until his retirement in 1996.

Wilson made important contributions to understanding the biological
basis of social behaviour. His book *The Insect Societies* (1971) detailed
his early research on behaviour and communication in social insect
communities, and argued that the same evolutionary forces have shaped
the behaviour patterns of insects and other animals, even including human
beings. Working with mathematician William Bossert, he also identified
the chemicals (pheromones) insects secrete to trigger social behaviours in
other individuals.

His 1975 book *Sociobiology: The New Synthesis* sparked controversy
when Wilson stressed the importance of genes in controlling a range of
human behaviours, including aggression, altruism and homosexuality. He
argued that the human mind is shaped by genetic inheritance as much as
it is by cultural influences, if not more, and that the influence of social and
environmental factors on human behaviour is limited.

STEPHEN JAY GOULD
10 SEPTEMBER 1941 – 20 MAY 2002

*American palaeontologist and evolutionary biologist,
famous for his popular science writing*

'We are the offspring of history, and must establish our own
paths in this most diverse and interesting of conceivable
universes – one indifferent to our own suffering, and
therefore offering us maximum freedom to thrive, or to fail,
in our own chosen way'

From Gould's *Wonderful Life* (2000)

Gould visited New York City's American Museum of Natural History when he was five years old. 'Awestruck' by the museum's fossilized *Tyrannosaurus rex* skeleton, he decided there and then to spend his life studying fossils. He did a first degree in geology and philosophy at Antioch College in Yellow Springs, Ohio, then a PhD in palaeontology back in New York at Columbia University. He then spent most of his career teaching at Harvard University and working at the New York museum that inspired his career.

Gould's greatest contribution to evolutionary biology was his theory of punctuated equilibrium, developed with American palaeontologist Niles Eldredge in 1972. In studies of trilobites (extinct marine arthropods), Eldredge noted that the fossil record shows sudden bursts of evolution with many new species appearing, followed by long periods of stability. **Darwin** had suggested this was just an artefact of an incomplete fossil record, and that evolution is smooth and continuous.

But Eldredge and Gould made the case that the episodic patterns in fossil record genuinely reflect life's history, and this view was widely accepted by the 1980s. (Some critics jokingly called the theory 'evolution by jerks', to which Gould responded that gradualism was 'evolution by creeps'.) Gould was a prolific writer and became world famous for his popular science books, including *Wonderful Life* (2000), which describes the unique Burgess Shale fossil field in the Canadian Rockies.

JAMES EPHRAIM LOVELOCK
26 July 1919 –

English scientist who developed the 'Gaia' hypothesis

'Is there any need to see Gaia, the only living planet in the
solar system? After all, despite the recent economic setback,
life continues to get better in most of the world ... Perhaps we
no longer need to see the Earth in reality when we can see it
so well on Google. It does matter, and it matters more than
any other thing: we have to see it as it really is because our
lives are wholly dependent upon the living Earth. We could
not survive for an instant on a dead planet like Mars, and we
need to understand the differences.'

Lovelock contemplates a future flight into space with Virgin Galactic
(*The Vanishing Face of Gaia*, 2009)

Lovelock was born in Letchworth Garden City,
Hertfordshire. He studied chemistry at Manchester University, graduating
in 1941, and earned a PhD in medicine from the London School of Hygiene
and Tropical Medicine in 1948. In 1959 he also received a DSc degree in
biophysics from London University. In 1961, he was appointed professor
of chemistry at Baylor University College of Medicine in Houston, Texas.

Since 1964, Lovelock has conducted an independent practice in science
and he has been a prolific inventor. He designed a sensitive detector
for traces of pesticides, which demonstrated widespread atmospheric
accumulation of chlorofluorocarbons.

But he is best known for introducing the Gaia theory, proposing
that all life on Earth and their environments are interlinked to form a
self-regulating system that maintains the conditions necessary for life
to thrive. Complicated interactions between life, the atmosphere and the
Earth's surface somehow stabilize oxygen levels in the atmosphere and
the Earth's surface temperature, for instance. During his career, Lovelock
has written many books and more than 200 scientific papers.

PETER MANSFIELD
9 October 1933 –

British physicist who developed magnetic resonance imaging for medical diagnosis

'We feel greatly honoured to be here and especially honoured to receive the awards for our work in magnetic resonance imaging (MRI). But there is another, darker side to our work ... the suffering and mental anguish associated with those afflicted with devastating illness. It is therefore all the more rewarding for us when someone takes a moment to write thanking us for the peace of mind felt by them in revealing their problem through MRI.'

From Mansfield's Nobel banquet speech (10 December 2003)

Mansfield left school at the age of 15 and worked as a printer's assistant in London. Following his national service, he studied for A-levels part time to qualify for university entrance. He then studied physics at Queen Mary College, University of London, from 1956 and earned a PhD there in 1962. Since 1964, he has worked at the University of Nottingham, where he was appointed full professor in 1979.

In the 1940s, **Bloch** and **Purcell** had described and measured nuclear magnetic resonance (NMR) in liquids and solids. NMR is an effect in which atomic nuclei placed in a strong magnetic field absorb and re-emit electromagnetic radiation at a specific resonance frequency which depends both on the strength of the magnetic field and the magnetic properties of the atomic isotope.

Mansfield showed that NMR could be used to produce images of the human body, describing how the tiny NMR radio signals from molecules in the body could be mathematically analysed and interpreted to create images of internal body tissues. His work also led to dramatic increases in the speed with which these images could be created. Since then, magnetic resonance imaging in medicine has radically improved diagnosis of disease. Mansfield was knighted in 1993 and he won a half share of the 2003 Nobel prize for physiology or medicine.

DONALD CARL JOHANSON
28 June 1943 –

American anthropologist who co-discovered the skeleton of the female australopithecine dubbed 'Lucy'

'Lucy is still a terribly important discovery all these years later ... Most of the evidence for human evolution older than three million years you could fit in the palm of your hand ... She showed us conclusively that upright walking and bipedalism preceded all of the other changes we'd normally consider being human, such as tool-making.'

From an interview with Johanson in *Time* online (4 March 2009)

Born to Swedish immigrant parents in Chicago, Illinois, Johanson studied chemistry and anthropology at the universities of Illinois and Chicago, where he earned a bachelor's degree in 1966. After completing a PhD at Chicago, he held a teaching post at Case Western Reserve University in Cleveland, Ohio. In 1981, he founded the Institute of Human Origins in Berkeley, California, now affiliated with Arizona State University in Tempe, where he is still director.

In 1974, Johanson found the fossilized remains of a female hominid the world came to know as Lucy (3.18 million years old) in the Afar region of Ethiopia. More than 40 per cent of the skeleton was preserved. Until the discovery of Lucy, scientists thought primates first evolved large brains and then began walking upright to free up their hands to use tools. But Lucy's skeleton and those of contemporary fossils from the site showed upright walking evolved before large brains and tool use – possibly to allow the hands to gather food for offspring.

Further study led Johanson to suggest in 1978 that Lucy belonged to a distinct new species, which he named *Australopithecus afarensis*. *A. afarensis* is now widely recognized as the ancestor of both *Australopithecus africanus* and modern man, *Homo sapiens*.

STEPHEN WILLIAM HAWKING
8 January 1942 –

Iconic physicist who has worked on the theory of black holes and the basic laws that govern the Universe

'My goal is simple – it is a complete understanding of the Universe, why it is as it is and why it exists at all'

Quoted in *Stephen Hawking's Universe* (1985) by John Boslough

 **Born in Oxford**, Hawking was brought up in London and St Albans. He studied physics at the University of Oxford and then did a PhD in cosmology at Cambridge University. Throughout his working life he has remained at Cambridge, where he held the post of Lucasian Professor of Mathematics, previously held by **Newton** and **Dirac**, from 1979 until 2009. Currently he is the director of research at the university's Centre for Theoretical Cosmology.

Hawking's work has clarified the basic laws that govern the Universe. With **Penrose**, he showed that **Einstein**'s general theory of relativity implied that space and time had an origin in the Big Bang, an extremely hot and dense state that rapidly expanded. Einstein's general relativity theory can't deal with such extreme conditions, and Hawking has played a leading role in the search for an alternative theory of 'quantum gravity' that would unify general relativity with quantum mechanics and allow scientists to understand the first moments of the Universe's creation.

Hawking has carried out revolutionary work on black holes, also superdense states of matter, which can form when a supermassive star explodes at the end of its life. Gravity can crush the remaining core to a point with such enormous density that nothing – not even light – can escape from a region surrounding it.

However, in 1974 Hawking predicted that, in fact, black holes are not completely black. Instead, quantum effects should make particles leak out from the black hole in a process that gradually accelerates as the black hole mass gets smaller until eventually it completely 'evaporates' in a puff of radiation. This process would be very slow, however, and therefore has not been observed. A black hole with a mass similar to that of the Sun or more would take more than 10^{67} years to evaporate – far longer than the current age of the Universe.

But Hawking also predicted that space could be littered with mini black holes left over after the Big Bang. In the extreme density of the hot fireball, dense lumps might collapse into black holes no bigger than a proton with relatively small masses – roughly the mass of a few cubic kilometres of rock (about 10 billion tonnes). In future, it might be possible to see them evaporate because they would be small enough to evaporate in the current era, creating bright flashes of radiation from space.

Hawking's life has been overshadowed by severe disability. Aged 21, he was diagnosed with motor neurone disease and doctors predicted then that he had no more than three years to live. He gradually lost the use of his arms and legs, as well as speech, and since 2009 he has been almost completely paralysed. He uses an electronic voice synthesizer to communicate.

Despite his disabilities, Hawking continues to combine family life (he has three children and three grandchildren) with research into theoretical physics as well as regular travel and public lectures. His popular science book *A Brief History of Time*, published in 1988, has sold more than 10 million copies and was on the London *Sunday Times* best-seller list for more than four years.

Hawking often raises the topic of God in his writing, famously stating in *A Brief History of Time* that if we could find a successful theory that explains everything in the Universe: 'It would be the ultimate triumph of human reason – for then we would know the mind of God'. In his 2010 book *The Grand Design* he adds: 'Because there is a law such as gravity, the Universe can and will create itself from nothing ... It is not necessary to invoke God to light the blue touch paper and set the Universe going.'

BURTON RICHTER

22 March 1931 –

American physicist who proved the existence of the 'charm' quark

'Nothing so strange and completely unexpected had happened in particle physics for many years ... The long-awaited discovery of anything which would give a clue to the proper direction in which to move in understanding the elementary particles loosed a flood of theoretical papers that washed over the journals in the next year.'

Richter describes the discovery of a new particle in his Nobel lecture
(11 December 1976)

 **Born in Brooklyn, New York**, Richter studied physics at the Massachusetts Institute of Technology, where he earned a first degree in 1952 and a PhD in 1956. He then worked at Stanford University in California, where he was appointed professor in 1967. He was director of the Stanford Linear Accelerator Center (SLAC) from 1984 to 1999.

At SLAC, Richter led the construction of a particle accelerator called SPEAR (Stanford Positron–Electron Asymmetric Ring). Using the accelerator, he led a team that in the mid-1970s discovered a new particle, now known as the J/ψ, which clarified the quark model of fundamental particles. By the 1960s, physicists had recognized that particles feeling the strong force such as protons and neutrons are made up of fractionally charged quarks that came in three 'flavours', labelled 'up', 'down' and 'strange'. However, inconsistencies in particle decay patterns led **Glashow** and others to suggest there must be a fourth 'charm' quark.

Richter's team confirmed this with the discovery of the J/ψ particle, a short-lived bound state of a charm quark and antiquark. **Ting**'s team discovered it independently at Brookhaven National Laboratory in New York State and for this work, Richter and Ting shared the 1976 Nobel prize for physics.

SAMUEL CHAO CHUNG TING
27 JANUARY 1936 –

Chinese–American physicist who proved the existence of the 'charm' quark

'The existence of J implies that we need at least four quarks to explain the phenomena observed so far. How many more quarks will we need if we find a new series of particles in higher energy regions? If we need a large family of quarks, are they the real fundamental blocks of nature?'

Ting speculates about the real number of fundamental particles in his Nobel lecture
(11 December 1976)

Ting was born to Chinese parents in Ann Arbor, Michigan. The family returned to China two months later and due to the Second World War he had no formal education until he was 12 years old. Ting moved to the US aged 20 to study maths and physics at the University of Michigan, where he earned a PhD in physics in 1962. Since 1969, he has been professor at the Massachusetts Institute of Technology.

In the mid 1970s, Ting's team discovered a new particle he called J at Brookhaven National Laboratory in New York state. This clarified the quark model of fundamental particles, proving the existence of a fourth known 'flavour' of quark called 'charm'. The particle is made of a charm quark and antiquark.

Richter discovered the particle independently and called it 'psi'; today it is usually called J/psi or J/ψ. For this discovery, Richter and Ting shared the 1976 Nobel prize for physics. Ting is currently the chief scientist for the Alpha Magnetic Spectrometer, a particle detector installed on the International Space Station on 19 May 2011. The experiment aims to detect cosmic ray particles, investigate antimatter and search for exotic particles new to science.

BENOÎT MANDELBROT
20 November 1924 – 14 October 2010

French–American mathematician considered the father of fractal geometry

'Not only are fractals beautiful, but they master key features of the roughness of nature and culture, including metal fractures, turbulence, financial markets and music ... The fractal geometry of roughness is set to expand rapidly and carve itself an increasingly central role.'

Mandelbrot quoted in *New Scientist* (18 November 2006)

Mandelbrot was born to a Lithuanian family in Warsaw, Poland, and the family moved to France when he was a child. From 1945 to 1947, he studied at the École Polytechnique in Paris. In 1949, he earned a master's degree in aeronautics at the California Institute of Technology in Pasadena, and then he returned to France where he earned a PhD in maths at the University of Paris in 1952. He spent most of his career at IBM's Thomas J. Watson Research Center in New York.

Mandelbrot studied a diverse range of mathematical fields, including economics and fluid dynamics, but he is best known for developing fractal geometry. He coined the term fractal in 1975 to describe objects that display self-similarity at different scales, so that magnifying them reveals small-scale details similar to the large-scale features. He also described one particular fractal now known as the Mandelbrot set, which is infinitely complex, despite being generated by a simple equation involving complex numbers (numbers that include an imaginary part involving the square root of negative one).

Mandelbrot used fractal geometry to explain many phenomena, such as how galaxies cluster, how brain structures develop and how crop prices change over time. He extensively popularized his work, writing several books including *The Fractal Geometry of Nature* (1982).

CARL EDWARD SAGAN
9 November 1934 – 20 December 1996

American astronomer and a prolific science communicator

'Today we have discovered a powerful and elegant way to
understand the Universe, a method called science; it has revealed
to us a Universe so ancient and so vast that human affairs seem
at first sight to be of little consequence'

From Sagan's book *Cosmos* (1980)

Sagan came from a Ukrainian Jewish family
living in New York. He studied at the University of
Chicago, where he earned four degrees including a
PhD in astronomy and astrophysics. He worked as
an astronomer at the Smithsonian Astrophysical
Observatory in Massachusetts and at Harvard
University and then Cornell, where he became a full
professor in 1971.

From the 1950s, Sagan was a consultant and adviser to NASA, where
one of his roles was to brief Apollo astronauts before their flights to the
Moon. He also played an active role in some of NASA's interplanetary
spacecraft programmes, including the Mariner, Viking, Voyager and
Galileo missions.

Sagan helped resolve several puzzles in planetary astronomy,
concluding that Venus's surprisingly high temperature is due to a runaway
greenhouse effect and that complex organic molecules explain the reddish
haze of Saturn's giant moon Titan. He was among the first to suggest that
Jupiter's moon Europa has a subsurface water ocean, which the Galileo
probe later indirectly confirmed.

But Sagan is most famous for his high-profile science communication,
including the 1980 television series *Cosmos: A Personal Voyage*, which he
co-wrote and narrated. He wrote many books, including *Cosmos* (1980),
which at the time became the best-selling science book on record. He also
vigorously promoted the search for extraterrestrial intelligence. His novel,
Contact, was made into a film starring Jodie Foster as a SETI researcher,
her character inspired by Sagan's colleague and friend **Tarter**.

(CLINTON) RICHARD DAWKINS
26 March 1941 –

English biologist and writer famous for his vocal opposition to religion

'It's been suggested that if the supernaturalists really had the powers they claim, they'd win the lottery every week. I prefer to point out that they could also win a Nobel prize for discovering fundamental physical forces hitherto unknown to science. Either way, why are they wasting their talents doing party turns on television? By all means let's be open-minded, but not so open-minded that our brains drop out.'

Dawkins mourns the 'epidemic of paranormal propaganda' on TV in his
Richard Dimbleby Lecture (1996)

Dawkins was raised by his English family in Nairobi, Kenya, where he lived until the age of eight, when his family returned to England. He studied zoology at the University of Oxford, earning his first degree in 1962 and his doctorate in 1966. After a spell at the University of California, Berkeley, he returned to Oxford where he has remained for the rest of his career, becoming the first Simonyi Professor for the Public Understanding of Science in 1995.

Dawkins has popularized the view that evolution operates at the level of the genes that control animal behaviour and development, not at the level of individual animals. In his 1976 classic book *The Selfish Gene*, he coined the term 'meme' (the behavioural equivalent of a gene), meaning an idea, behaviour or style that spreads from person to person within a culture.

He is also famously a vocal critic of creationism, the notion that the Universe and living organisms originated by divine purpose without evolution. He vigorously criticizes pseudoscience and alternative medicine. He has also campaigned against human overpopulation, arguing that Roman Catholic promotion of 'natural' methods of contraception will ironically limit population ultimately through starvation.

CARL RICHARD WOESE
15 July 1928 –

*American molecular biologist who identified the archaea
as a new domain of life*

'It wasn't long before I realized, "Oh my God, we've made
a mistake here in assuming that all prokaryotes were the
same kind." Because what a prokaryote was, by definition,
was a little tiny bacterial looking thing. But this just didn't
fit with its signature ... It just didn't fit the signature of the
bacteria or the signature of the eukaryotes. So that was really
a thrilling moment.'

Woese describes his discovery of a third domain of life

Woese was born in Syracuse, New York, and studied maths
and physics at Amherst College in Massachusetts, graduating in 1950. He
then did a doctorate in biophysics at Yale University. In 1964, he joined
the faculty of the University of Illinois at Urbana-Champaign where he is
still professor today.

Woese is best known for proposing in 1977 a new domain of life.
Previously, life had been classified into two types, essentially bacteria and
'everything else'. But through rigorous genetic analysis, Woese recognized
that some single-celled life previously lumped in with bacteria actually
have an independent evolutionary history and deserve their own branch in
the tree of life – the archaea.

Although they are usually similar to bacteria in size and shape,
archaea possess genes and some metabolic pathways that are more
closely related to those of eukaryotes (plants and animals with DNA
in a cell nucleus). Archaea can also use a surprisingly large variety of
energy sources, including ammonia, metal ions and hydrogen gas. They
have since been found in many habitats including oceans, soils and the
human digestive tract.

ALAN HARVEY GUTH

27 FEBRUARY 1947 –

American physicist who proposed the theory of the inflationary Universe

'Despite its name, the Big Bang theory is not really a theory of a bang at all. It is really only a theory of the aftermath of a bang ... The standard Big Bang theory says nothing about what banged, why it banged, or what happened before it banged. The inflationary Universe is a theory of the "bang" of the Big Bang.'

From Guth's *The Inflationary Universe* (1998 edition)

Born in New Brunswick, New Jersey, Guth studied physics at the Massachusetts Institute of Technology (MIT), where he earned a PhD in 1972. He held posts at Princeton University, Columbia and Cornell before in 1980 becoming an associate professor at MIT, where he is currently Jerrold Zacharias Professor of Physics.

In 1979, while on a sabbatical at Stanford University, Guth proposed a radical theory that has completely changed our picture of the Universe's origins. At the time, no theory explained some features of the cosmos, such as its smoothness and the 'flatness' of space (light travels in straight lines); it seemed more likely that space would have had some curvature.

Guth resolved these problems by showing that the Universe originated in a state called a 'false vacuum' that was destined to decay and trigger exponential expansion. According to this 'inflation' theory, the Universe doubled in size about once every 10^{-35} seconds during the first split second after the Big Bang. So the Universe ballooned to enormous size and therefore space seems flat, just as the spherical Earth seems flat to someone walking on its surface. The theory neatly accounts for many other features of the cosmos as well.

STANLEY BEN PRUSINER
28 May 1942 –

American neurologist and biochemist who discovered prions, faulty proteins that cause 'mad cow disease' and its human form CJD

'Because our results were so novel, my colleagues and I had great difficulty convincing other scientists of the veracity of our findings and communicating to lay people the importance of work that seemed so esoteric! As more and more compelling data accumulated, many scientists became convinced. But it was the "mad cow" epidemic in Britain and the likely transmission of bovine prions to humans producing a fatal brain illness called Creutzfeldt-Jakob disease that introduced prions to the public.'

From Prusiner's Nobel banquet speech (10 December 1997)

Prusiner studied chemistry and medicine at the University of Pennsylvania. In 1972, during a residency at the University of California, San Francisco, he admitted a patient dying from Creutzfeldt-Jakob disease (CJD), suspected at the time to be caused by a mysterious 'slow virus'. Intrigued by this pathogen's strange properties, as well as by similarities between CJD and other rare diseases such as scrapie in sheep, Prusiner decided to research its origins.

He showed that the agent causing scrapie contains protein but none of the nucleic acids characteristic of a virus's genetic material. His controversial conclusion, published in 1982, was that the diseases are caused by infectious proteins he called 'prions'. Later analysis of their structures has led to widespread acceptance that scrapie, CJD and bovine spongiform encephalopathy ('mad cow disease') are caused by proteins that have folded into the wrong shape and make other proteins morph into the same faulty state.

Prion diseases destroy brain tissue and are always fatal. For his work on their origins, Prusiner won the 1997 Nobel prize for physiology or medicine.

LUC ANTOINE MONTAGNIER
18 August 1932 –

French virologist who discovered HIV, the virus that causes AIDS

'By September 1983, I was able to make a synthesized presentation of all our data favouring a causal link between the virus and the disease ... This presentation was received with scepticism ... This situation is not infrequent in science, since new discoveries often raise controversy. The only problem is that it was a matter of life and death for blood-transfused people.'

From *Les Prix Nobel 2008* (Nobel Foundation, 2009)

Born in Chabris, near Tours, Montagnier studied medicine and sciences at the University of Poitiers, where he earned his first degree in 1953. He earned a PhD in medicine at the University of Paris in 1960. After holding several posts in Britain and France, he joined the Pasteur Institute in Paris, where he was director of the Viral Oncology Unit from 1972 to 2000.

In 1982, Montagnier began work on AIDS (acquired immunodeficiency syndrome) when some scientists began to suspect that it was caused by a virus, possibly a 'retrovirus' that inserts its genetic material into the DNA of host cells. Patients infected by a retrovirus called HTLV were already known to sometimes develop a wasting syndrome, possibly as a result of immune suppression. Montagnier's team showed that a different retrovirus, now called HIV, causes AIDS. It is from a subclass of retroviruses that doesn't cause immunodeficiency in animals, a surprising result that delayed acceptance of the HIV–AIDS link.

For this work, Montagnier won a share of the 2008 Nobel prize for physiology or medicine. In 2009, he was widely criticized for announcing the detection of electromagnetic signals from DNA in highly diluted water, a result that homeopathy proponents claimed provided scientific evidence for their therapies.

WALTER ALVAREZ
3 October 1940 –

American geologist who promoted the idea that a devastating asteroid or comet impact was responsible for the extinction of the dinosaurs

'The story of the impact and the extinction it caused is dramatic and horrifying, although the passage of 65 million years makes it seem comfortably remote ... It is a story of challenge by a few geologists, of conflict and friendship, of adventure in distant places.'

From Alvarez's book *T. Rex and the Crater of Doom* (1997)

The son of Luis Alvarez, Walter Alvarez was born in Berkeley, California. He studied geology at Carleton College in Minnesota and Princeton University in New Jersey, where he earned a PhD in 1967. He worked for an oil company in the Netherlands and Libya, then did geological dating work in Italy before in 1977 taking a post at the University of California, Berkeley, where he worked for the rest of his career.

At Berkeley, he began studying the mass extinction of life at the end of the Cretaceous period 65 million years ago, when around 75 per cent of all species on Earth became extinct. Working with his father and others, Walter Alvarez showed that sedimentary layers around the world dating to the time of the extinction contain high concentrations of the element iridium, which is rare in the Earth's crust.

But iridium is abundant in asteroids and comets. So Alvarez's team proposed that a massive asteroid or comet hit the Earth at the end of the Cretaceous period. The impact would have dramatically altered the Earth's climate by raising sunlight-blocking dust into the atmosphere globally. The idea remained controversial, however, until 1991, when geophysicist Glen Penfield discovered the scene of the crime – the Chicxulub crater, a vast depression off the Yucatán peninsula in the Gulf of Mexico, which dates to an impact 65 million years ago.

ALEC JOHN JEFFREYS
9 January 1950 –

*British geneticist who developed DNA
fingerprinting techniques*

'My life changed on Monday morning at 9.05 am, 10 September
1984. What emerged was the world's first genetic fingerprint. In
science it is unusual to have such a "eureka" moment. We were
getting extraordinarily variable patterns of DNA, including from
our technician and her mother and father, as well as from non
human samples. My first reaction to the results was "this is too
complicated", and then the penny dropped and I realized we had
genetic fingerprinting.'

From an interview with Jeffreys

 **Jeffreys studied biochemistry** and genetics
at Oxford University, graduating with a first degree
in 1972 and a doctorate in 1975. After that he briefly
worked on mammalian genetics at the University
of Amsterdam before moving to the University of
Leicester, where he has remained throughout his
career, as professor from 1987.

At Leicester, Jeffreys's work has focused on
exploring human DNA diversity and the mutation processes that generate
it as genetic information is passed from parent to child. But he is best
known for the invention of DNA fingerprinting in 1984. In this technique,
enzymes cut DNA into short pieces that are separated according to size on
a gel. Comparing the DNA fingerprint of one individual to that of another
individual reveals whether or not they are closely related.

These techniques are now widely used to resolve paternity and
immigration disputes. From 1985, Jeffreys's team developed a variation
called genetic profiling to quickly establish whether a DNA sample
matches that of a given person, a technique that has been refined for
widespread use in forensic investigations. He was knighted in 1994.

BARRY JAMES MARSHALL
30 September 1951 –

Australian medical researcher who proved that bacteria cause peptic ulcers

'While it is true that MacFarlane Burnet injected himself with the rabbit myxoma virus, and I did actually infect myself with *Helicobacter pylori*, I don't suggest to other aspiring Aussie scientists that this process will guarantee a Nobel prize. But to young people listening tonight I would say, find passion in your work – whatever it is ... Work hard, keep balance in your life and, just in case, always be nice to Swedish people.'

From Marshall's Nobel banquet speech (10 December 2005)

Marshall was born in Kalgoorlie in Western Australia. He studied medicine at the University of Western Australia, where he graduated with a bachelor's degree in medicine and surgery in 1975. After holding posts at the Royal Perth Hospital, he became professor of medicine at the University of Virginia in 1986, and in 1997, he returned to the University of Western Australia as professor.

Marshall and his colleague Robin Warren overturned the conventional wisdom that stress and diet are major causes of peptic ulcer disease. Warren had noted that the stomachs of patients with gastritis (inflammation of the stomach) were often colonized by small curved bacteria. Marshall cultivated these previously unknown bacteria (later called *Helicobacter pylori*) and together they showed the bugs are present in almost all patients with gastric inflammation or peptic ulcers. During the work, Marshall deliberately drank a culture of *H. pylori* expecting eventually to develop an ulcer, but he developed gastritis after just five days.

Antibiotics can now cure these conditions, so that peptic ulcer is no longer a chronic, frequently disabling condition. For their discovery, Marshall and Warren shared the 2005 Nobel prize for physiology or medicine.

KARY BANKS MULLIS
28 December 1944 –

American chemist who invented a powerful way to amplify DNA

'I had solved the most annoying problems in DNA chemistry in a single lightning bolt ... I could make as much of a DNA sequence as I wanted and I could make it on a fragment of a specific size that I could distinguish easily. Somehow, I thought, it had to be an illusion. Otherwise it would change DNA chemistry forever. Otherwise it would make me famous. It was too easy.'

Mullis remembers his 'eureka' moment in his Nobel lecture (8 December 1993)

Born in Lenoir, North Carolina, Mullis studied chemistry at the Georgia Institute of Technology and earned a PhD in biochemistry from the University of California, Berkeley, in 1972. The following year, he began researching paediatric cardiology at the University of Kansas Medical School and in 1979 he joined the Cetus Corporation in Emeryville, California, as a DNA chemist. Since 1986 he has worked for many commercial companies.

At Cetus, Mullis invented the polymerase chain reaction (PCR) technique. A method of amplifying DNA, PCR multiplies a single, microscopic strand of the genetic material billions of times within just a few hours. The process has found wide-reaching applications in forensics, medicine, biotechnology and genetics. It also allows for DNA to be extracted from fossils and opened up the new scientific field of palaeobiology. For inventing PCR, Mullis won a half share of the 1993 Nobel prize for chemistry.

Mullis has developed a reputation as a spirited and unconventional freethinker. His inventions include a technique that slashed the amount of time it takes to filter DNA from blood as well as an ultraviolet-sensitive plastic that changes colour in response to light. He published the autobiographical *Dancing Naked in the Mind Field* in 1998.

STEVEN CHU
28 February 1948 –

*American physicist who developed a way to cool
and trap atoms using laser light*

'The trap worked. We could actually see the random walk loading
with our own eyes. A tiny dot of light grew in brightness as more
atoms fell into the trap. During the first days of trapping success,
I ran up and down the halls, pulling people into our lab to share
in the excitement.'

From Chu's Nobel lecture (8 December 1997)

Born to a Chinese immigrant family in
St Louis, Missouri, Chu studied physics and maths
at the University of Rochester in New York State.
In 1976, he earned a PhD from the University of
California, Berkeley, before joining Bell Laboratories
in New Jersey in 1978. In 1987, he moved to
Stanford University in California. In 2004, Chu
became director of the Lawrence Berkeley National
Laboratory and President Barack Obama appointed him as US Secretary
of Energy in 2009.

While at Bell labs, Chu led a team that invented a way to cool and trap
atoms using light. The technique involves using several lasers tuned to
specific frequencies as 'optical molasses', so that whichever way an atom
tries to move, it experiences an opposing force. The atoms then move
so slowly that they have an effective temperature of nearly absolute zero
(−273.15°C). Laser traps have played a vital role in clarifying atomic physics
and testing quantum mechanics, and they could allow the development of
amazingly accurate atomic clocks.

For this work, Chu won a third share of the 1997 Nobel prize for
physics. He has also been a vigorous campaigner for action to alleviate
global warming, advocating a shift from fossil fuel burning to nuclear
power and alternative energy.

FRANCIS SELLERS COLLINS
14 APRIL 1950 –

American physician and geneticist who uncovered many genes linked to disease

'We are now in the midst of a genetic revolution that will touch all of us in numerous ways: this revolution involves common diseases like diabetes, heart disease, cancer, asthma, arthritis, Alzheimer's disease, and more; mental health and personality; decisions about child bearing; and even our ethnic histories. We now see that the language spoken by our DNA is the language of life itself.'

From Collins's book *The Language of Life: DNA and the Revolution in Personalised Medicine* (2010)

Collins trained in chemistry at the University of Virginia in Charlottesville and Yale University in Connecticut before studying medicine at the University of North Carolina in Chapel Hill. After postgraduate training, he held several professorships at the University of Michigan, Ann Arbor. In 1993, he succeeded **Watson** as director of the National Human Genome Research Institute in Bethesda, Maryland. In 2009, President Barack Obama nominated him to be director of the National Institutes of Health.

Since the 1980s, Collins has developed ways to identify disease genes in DNA. His technique of 'positional cloning' became the first method of identifying disease genes even when little or no information about the biochemical basis of the disease is available. His research teams have identified the genes for many diseases including cystic fibrosis and Huntington's disease.

From 1993, Collins oversaw the international Human Genome Project to identify all the 20,000 to 25,000 protein-coding genes in human DNA. He has been a vocal supporter for keeping genetic information private and has campaigned to prevent the insurance industry using it for discrimination. Collins is also an evangelical Christian; he published *The Language of God: A Scientist Presents Evidence for Belief* in 2006.

DAVID ELIESER DEUTSCH
1953 –

*Israeli-born physicist who has pioneered
quantum computation*

'Edison said that research is 1 per cent inspiration and
99 per cent perspiration – but that is misleading, because
people can apply creativity even to tasks that computers
and other machines do uncreatively. So science is not
mindless toil for which rare moments of discovery are the
compensation: the toil can be creative, and fun, just as the
discovery of new explanations is.'

From Deutsch's book *The Beginning of Infinity: Explanations that Transform the World* (2011)

Born in Haifa, Deutsch was educated at the universities of Cambridge
and Oxford. He worked for several years at the University of Texas at
Austin, before returning to Oxford, where he has been a non-stipendiary
Visiting Professor of Physics since 1999.

Deutsch pioneered the field of quantum computers by formulating
a description for a quantum generalization of the universal 'Turing
machine'. In the 1930s, Turing outlined the concept of a classical Turing
machine, which describes how all current computers work – a sequence of
steps operates on an input to produce an output. Deutsch showed in 1985
that quantum computers using information represented by the quantum
states of atoms (for instance, their spin states) would not be bound by the
same rules because the information is encoded as a superposition of two
or more states at once.

Deutsch has also clarified the physical processes needed to implement
quantum computation in the laboratory, which is still in an early research
phase but in theory could lead to ultrafast computation. He supports the
many-worlds interpretation of quantum mechanics, which asserts that all
our possible futures play out in myriad parallel universes. In 1998, he
was awarded the Dirac Prize and Medal of the Institute of Physics and in
2005, he won the $100,000 Edge of Computation Prize established by
Edge Foundation with funding from American financier Jeffrey Epstein.

HAROLD (HARRY) WALTER KROTO
7 OCTOBER 1939 –

British chemist who co-discovered buckyballs

'My advice is to do something which interests you or which
you enjoy (though I am not sure about the definition of
enjoyment) and do it to the absolute best of your ability. If it
interests you, however mundane it might seem on the surface,
still explore it because something unexpected often turns
up just when you least expect it. With this recipe, whatever
your limitations, you will almost certainly still do better than
anyone else. Having chosen something worth doing, never
give up and try not to let anyone down.'

Kroto's advice to students (from his Nobel autobiography, *Les Prix Nobel 1996*,
Nobel Foundation, 1997)

**Born to first-generation German
parents** in the town of Wisbech, Cambridgeshire,
Kroto studied chemistry at the University of
Sheffield, where he earned his first degree in 1961
and a PhD in 1964. In 1967, he joined the faculty of
the University of Sussex in Brighton, where he was
appointed full professor in 1985, and since 2004 he
has been professor at Florida State University.

Working with Robert Curl, Richard Smalley and other scientists from
Rice University in Texas, Kroto discovered the soccer-ball-shaped carbon
molecule C_{60} in 1985 after using a laser to vaporize graphite. The molecules
are called buckminsterfullerene (popularly known as 'buckyballs') after
the architect Richard Buckminster Fuller, whose geodesic domes they
resemble. Buckyballs and related molecules have since been found in
space, and their discovery has opened up new directions in nanotechnology.

For this work, Kroto, Curl and Smalley shared the 1996 Nobel prize for
chemistry. Kroto was knighted in the same year.

JAMES HANSEN
29 March 1941 –

American climate scientist and environmental campaigner

'Continued exploitation of all fossil fuels on Earth threatens not only
the other millions of species on the planet but also the survival of
humanity itself – and the timetable is shorter than we thought'

From Hansen's book *Storms of my Grandchildren* (2009)

Hansen was born in Denison, Iowa, and studied physics and
astronomy at the University of Iowa, where he graduated with a BA in
1963 and a PhD in 1967. He became a NASA graduate trainee and studied
at the universities of Kyoto and Tokyo in Japan but has throughout his
career been based at NASA's Goddard Institute for Space Studies, which
he now heads. He is also adjunct professor of Earth and environmental
science at Columbia University.

Hansen's early research focused on radiative energy transfer models
for Venus's atmosphere. He later refined these models for application
to the Earth's atmosphere and investigated the effects of trace gases
and aerosols on the Earth's climate. He has argued that anthropogenic
influences on the Earth's climate are now more powerful than natural ones,
emphasizing that they could cause dramatic climate change depending on
levels of stability of the Greenland and Antarctic ice sheets.

His 1988 testimony on climate change to US congressional
committees raised awareness of the problem of global warming. A
committed activist calling for action to mitigate the effects of climate
change, he has been arrested several times during protests and
claimed to have been muzzled on the issue by the George W. Bush
administration. In 2009, he published *Storms of My Grandchildren*,
which makes the case that our planet is hurtling even more rapidly than
people realize to a climatic point of no return.

TIM BERNERS-LEE
8 June 1955 –

British computer scientist credited with inventing the World Wide Web

'We want the Web to reflect a vision of the world where everything is done democratically, where we have an informed electorate and accountable officials. To do that we get computers to talk with each other in such a way as to promote that ideal.'

Quote from an interview with Berners-Lee in *IT Now* (March 2006)

Berners-Lee was born in London and studied physics at Oxford University from 1973 to 1976. After that he worked for commercial technology companies in Dorset. In 1980, while working as an independent contractor at CERN, the European centre for particle physics on the French–Swiss border, he wrote a program called 'Enquire' for storing information using random associations. The system was based on hypertext (references displayed on a computer that a user can immediately access, for instance by clicking a mouse).

After working for a computer company in Bournemouth, Berners-Lee returned to CERN in 1984. In 1989, he proposed a global hypertext project – the World Wide Web. This first became available within CERN in December 1990 and on the Internet overall in the summer of 1991. Since then, he has continued revising its design in response to user feedback.

In 1994, Berners-Lee founded the World Wide Web Consortium, a Web standards organization with offices at the Massachusetts Institute of Technology and around the world to develop software and tools to lead the Web to its full potential. In December 2004, he became professor of computer science at the University of Southampton in England.

He is also a director of the World Wide Web Foundation, launched in 2009 to fund and coordinate efforts to further the potential of the Web to benefit humanity. He was knighted in 2004.

EDWARD WITTEN
26 August 1951 –

American theoretical physicist who has developed superstring and quantum gravity theories

'Supersymmetry is the framework in which theoretical physicists have sought to answer some of the questions left open by the Standard Model of particle physics. The Standard Model, for example, does not explain the particle masses. If particles had the huge masses allowed by the Standard Model, the Universe would be a completely different place. There would be no stars, planets, or people, since any collection of more than a handful of elementary particles would collapse into a black hole.'

From Witten's foreword to *Supersymmetry: Unveiling the Ultimate Laws of Nature* by Gordon Kane (2001 edition)

 **Born in Baltimore, Maryland,** Witten studied history and linguistics at Brandeis University in Waltham, Massachusetts, then briefly studied economics before enrolling in applied maths at Princeton University in New Jersey. He earned a PhD in physics in 1976 and was appointed professor at Princeton in 1980. Since 1987, he has been professor at the Institute for Advanced Study, also in Princeton.

Witten has made extensive contributions to theories of nature's fundamental forces and matter. For instance, in the 1990s he developed 'M-theory', which unifies various different string theories. String theories portray all the particles and all forms of energy in the Universe as hypothetical one-dimensional 'strings', tiny building blocks that have length but no height or width. M-theory shows how these theories complement each other if strings are really just one-dimensional slices of a membrane that vibrates in an 11-dimensional space.

Witten has written more than 300 scientific papers. In 1990, he was awarded the Fields Medal by the International Mathematical Union, becoming the first physicist to win this top mathematical honour.

GEORGE SMOOT
20 FEBRUARY 1945 –

*American astrophysicist who led a team that discovered
ripples in the cosmic microwave background*

'I compared looking for the cosmic background radiation to
"listening for a whisper during a noisy beach party while radios blare,
waves crash, people yell, dogs bark and dune buggies roar"'

From Smoot's Nobel autobiography

Smoot studied for a degree in maths and physics and then
earned a PhD in particle physics at the Massachusetts Institute of
Technology in 1970. During his PhD, he heard about **Penzias** and **Robert
Wilson**'s discovery of the cosmic microwave background. This radiation
started flooding through the Universe when it first became transparent to
light, 380,000 years after the Big Bang.

Smoot then moved to the University of California at Berkeley and
the Lawrence Berkeley National Laboratory, where he worked with **Luis
Alvarez** to develop a microwave detector to study the background radiation
from a high-flying U-2 spy plane. Their measurements showed the
radiation has its highest frequency in the direction towards which our
galaxy is moving due to the **Doppler** effect. That implied the Milky Way
moves at an enormous speed due to the gravitational pull of distant galaxy
clusters; in a smooth Universe, there shouldn't be such a strong pull from
one direction. So this suggested the Universe must have been lumpy since
early times and the cosmic radiation itself should look uneven, containing
tiny ripples in wavelength.

With John Mather from NASA's Goddard Space Flight Center in
Maryland, Smoot led NASA's COBE (Cosmic Background Explorer)
satellite mission to study the microwave background in unprecedented
detail. Sure enough, it detected the telltale ripples in 1991. Since then,
studies of the microwave background ripples have clarified the history of
the Universe and pinned its age down to 13.7 billion years. Smoot and
Mather shared the 2006 Nobel prize for physics.

JILL CORNELL TARTER
16 January 1944 –

American astronomer who leads the search for
extraterrestrial intelligence (SETI)

'For many millennia, humans have been on a journey to
find answers, answers to questions about naturalism and
transcendence, about who we are and why we are, and of
course, who else might be out there. Is it really just us? Are
we alone in this vast Universe of energy and matter and
chemistry and physics? Well, if we are, it's an
awful waste of space.'

From Tarter's talk to a 2009 TED (Technology, Entertainment, Design) conference

Born in upstate New York, Tarter studied engineering physics
at Cornell University and did a PhD in astronomy at the University of
California, Berkeley. She is currently professor at the SETI Institute in
Mountain View, California, where she is also director of the Center for
SETI Research.

In the early 1990s, Tarter was project scientist for NASA's SETI
programme, the High Resolution Microwave Survey, and she has conducted
many observational programmes at radio observatories worldwide. Since
NASA stopped funding SETI research in 1993, she has vigorously led
efforts to secure private funding for it.

She is a member of the management board for the Allen Telescope
Array, a joint project between the SETI Institute and the University of
California at Berkeley to build an array of up to 350 large radio antennas
dedicated to astronomical observations and a simultaneous search for
artificial-looking radio signals from intelligent extraterrestrial civilizations.
However, funding problems have made the future of the project uncertain.

A friend and colleague of **Sagan**, Tarter was portrayed as a fictional
character who detects extraterrestrial signals in Sagan's novel *Contact*.
Jodie Foster played her character in the film version.

Indian neuroscientist whose experiments have generated new ideas about how our brains work

'How can a three-pound mass of jelly that you can hold in your palm imagine angels, contemplate the meaning of infinity, and even question its own place in the cosmos? ... With the arrival of humans, it has been said, the Universe has suddenly become conscious of itself. This, truly, is the greatest mystery of all.'

From Ramachandran's *The Tell-Tale Brain* (2010)

Born in Tamil Nadu, India, Ramachandran initially trained as a doctor in Madras and obtained a PhD from the University of Cambridge. He was appointed assistant professor of psychology at the University of California, San Diego, in 1983 and has been a full professor there since 1998. He is currently director of the university's Center for Brain and Cognition and adjunct professor of biology at the Salk Institute in La Jolla.

Ramachandran is best known for simple but revealing experiments in behavioural neurology that have had a profound impact on the way we think about the brain. He has studied synaesthesia, a condition in which people experience mixed-up sensations – for example, different musical notes might always evoke the same distinct colours. Ramachandran's experiments characterized the condition in detail and revealed that it is a genuine sensory effect rather than some kind of memory association.

He has also put forward theories to explain phantom limb sensations, in which people commonly perceive that an amputated or missing limb is still attached to their body and causing sensations like pain or tingling. He also showed that introducing visual feedback from mirrors can be an effective way of treating paralysis and chronic pain, including phantom pain. In 2011, *Time* magazine named Ramachandran as one of the 100 most influential people in the world. His popular science books include *Phantoms in the Brain* (1998) and *The Tell-Tale Brain* (2010). **Dawkins** has called him the 'Marco Polo of neuroscience'.

ANDREW JOHN WILES
11 April 1953 –

British mathematician who proved Fermat's Last Theorem

'I found this one book, which was all about one particular
problem – Fermat's Last Theorem. This problem had been
unsolved by mathematicians for 300 years ... Here was
a problem, that I, a ten year old, could understand, and I
knew from that moment that I would never let it go.
I had to solve it.'

From an interview with Wiles for NOVA programme
The Proof's website

Wiles was born in Cambridge and studied maths at Oxford
University, graduating in 1974. He then went to Cambridge University to
do a PhD and he was appointed as assistant professor at Harvard University
in 1977. He became professor at Princeton University in 1982. In 2011, he
returned to Oxford University as Royal Society Research Professor.

When he was ten years old, Wiles came across Fermat's Last Theorem,
a long-standing problem in mathematics. The 17th-century French lawyer
and mathematician Pierre de Fermat claimed to have proven that the
equation $x^n + y^n = z^n$ has no non-zero whole-number solutions when n is
an integer greater than 2. (So for example, you can't find any numbers x, y
and z such that $x^3 + y^3 = z^3$.)

However, Fermat didn't write down the proof he claimed to have.
After working on the problem for seven years, Wiles finally produced a
100-page mathematical proof that Fermat was correct and he presented
it at a Cambridge conference in 1993. It later turned out that there was a
worrying gap in the proof, but Wiles managed to circumvent the problem
in 1994. For proving the theorem, Wiles was knighted in 2000.

IAN WILMUT
7 July 1944 –

English embryologist best known as the leader of a team that created the first cloned mammals including Dolly the sheep

'Dolly was a bonus ... Sometimes when scientists work hard, they also get lucky, and that's what happened.'

Wilmut quoted in *Time* magazine (December 1997)

 **Wilmut was born in Hampton Lucy**, a village in Warwickshire. He studied agriculture at the University of Nottingham but his interests shifted towards animal science. At the University of Cambridge, he was awarded a PhD in 1971 for research on preserving boar semen by freezing it. In further research on animal reproduction at Cambridge, Wilmut and his colleagues produced the first calf, named Frosty, from a frozen embryo.

In 1974, Wilmut joined the Animal Breeding Research Station near Edinburgh, which later changed its name to the Roslin Institute. Currently he is director of the Medical Research Council Centre for Regenerative Medicine at the University of Edinburgh, which aims to develop new treatments for human diseases using stem cells. These 'blank canvas' cells can be coaxed into developing into a diverse range of tissues, such as liver or nerve cells, and could one day be used to treat a wide range of diseases resulting from tissue damage. Wilmut was knighted in 2008.

His research has focused on eggs and sperm cells, as well as the factors controlling normal or abnormal development of embryos. In 1986, he began a project to genetically modify female sheep so that they produced proteins needed to treat human disease in their milk. The starting point was to inject a few hundred copies of the desired gene into the cell nucleus of an early sheep embryo. However, this turned out to be frustratingly inefficient – thousands of fertilized eggs had to be injected to get just one animal that was born and developed normally to adulthood with the therapeutic gene active in the correct tissues.

To resolve this, Wilmut's team wanted to manipulate the genes of embryonic sheep cells cultured in a flask, in which they could insert a gene, allow the cells to multiply and check whether the genetic manipulation had worked. If so, the flask would contain thousands of correctly programmed cells to experiment with. But that meant finding a way to create living sheep from these engineered cells.

To do this, they transferred the successfully engineered nuclei of lab-grown cells from sheep embryos and foetuses into eggs that had had their own genetic material removed. These developed into embryos that were implanted into a surrogate mother sheep who gave birth to the first cloned mammals, Megan and Morag, in 1995.

More sensationally, Wilmut's team achieved the same feat using cells from an adult sheep's mammary gland cultured in a flask. In 1996, a surrogate mother sheep gave birth to Dolly (named after country singer Dolly Parton) – a genetic carbon copy of the adult female that donated the mammary cells. Dolly lived until 2003. Since Dolly, researchers have cloned many large and small mammals including horses, goats, cows, mice, pigs, cats and rabbits.

In 2005, Wilmut was granted a license to clone human embryos for the purpose of culturing stem cells. Scientists hope that treatments using cells derived from human embryos, including neurons for spinal cord repair, could one day revolutionize medicine. They are still in the early trial phase. Eventually, it might be possible to take adult stem cells from a patient needing treatment and programme them to return to an embryonic-like state. These 'pluripotent' stem cells could then diversify into any tissue the patient needs and be transplanted into the patient without any risk of tissue rejection by the immune system.

Controversy over Dolly surfaced in 2008 when some former employees of the Roslin Institute, who were not directly involved with the work, argued that Wilmut has claimed undue credit for creating Dolly and petitioned the Queen to strip him of his knighthood. In 2007, Wilmut had admitted that he played a lesser role in the breakthrough than his Roslin colleague Keith Campbell.

NEIL DEGRASSE TYSON
5 OCTOBER 1958 –

American astrophysicist and science communicator

'Our species is dumber than we normally admit to ourselves ... Chimpanzees are an evolutionary hair's-width from us yet we can agree that no amount of tutelage will ever leave a chimp fluent in trigonometry. Now imagine a species on Earth, or anywhere else, as smart compared with humans as humans are compared with chimpanzees. How much of the Universe might they figure out?'

From Tyson's *Death by Black Hole* (2007)

Tyson studied physics at Harvard University, where he earned a first degree in 1980. He later earned a PhD in astrophysics from Columbia University. Since then he has held numerous positions at institutions including the University of Maryland, Princeton University in New Jersey and the American Museum of Natural History in New York. He is currently director of the museum's Hayden Planetarium.

Tyson has a broad range of professional research interests, including the physics of star formation, exploding stars, dwarf galaxies and the structure of our galaxy, the Milky Way. But he is best known for his prolific science communication efforts. He makes regular appearances on popular science TV series and has written many books including *Death by Black Hole and Other Cosmic Quandaries* (2007), which appeared on *The New York Times* bestseller list. In 2000, *People* magazine bestowed on him the curious accolade of 'sexiest astrophysicist alive'.

President George W. Bush appointed Tyson to serve on the Commission on the Future of the United States Aerospace Industry in 2001. The committee's final report was published in 2002 and detailed recommendations to promote a thriving future of transportation, space exploration and national security. In 2004, he was also appointed to serve on the President's Commission on Implementation of United States Space Exploration Policy (popularly known as the 'Moon, Mars and Beyond' commission).

BRIAN GREENE

9 February 1963 –

American theoretical physicist who has pioneered superstring theory

'The equations of general relativity and quantum mechanics, when combined, begin to shake, rattle, and gush with steam like a red-lined automobile. Put less figuratively, well-posed physical questions elicit nonsensical answers from the unhappy amalgam of these two theories ... Can it really be that the Universe at its most fundamental level is divided, requiring one set of laws when things are large and a different, incompatible set when things are small? Superstring theory, a young upstart compared with the venerable edifices of quantum mechanics and general relativity, answers with a resounding no.'

From Greene's *The Elegant Universe* (2005 edition)

Greene was born in New York City and studied physics at Harvard University, where he graduated with his first degree in 1984. After that he became a Rhodes Scholar at Oxford University, where he earned a PhD in 1986. In 1990 he joined the physics faculty of Cornell University, where he became professor in 1995. Since 1996, he has been professor of physics and maths at Columbia University.

Greene is renowned for his groundbreaking discoveries in superstring theory, which aims to create a quantum theory of gravity and a unified theory of all the forces and matter in nature. If successful, this would realize **Einstein**'s dream of a single, all-encompassing theory of the Universe. Much of Greene's research has focused on the implications of the extra space dimensions that superstring theory requires. He has shown that unlike in Einstein's general theory of relativity, superstring theory predicts that the fabric of space can rip apart.

Greene is also well known for his popular science writing. His first book, *The Elegant Universe* (1999), sold more than a million copies worldwide.

(JOHN) CRAIG VENTER
14 October 1946 –

American biologist who sequenced the human genome and created the first cell with a synthetic genome

'It's about scientific power. The real problem is that the understanding of science in our society is so shallow. In the future, if we want to have enough water, enough food and enough energy without totally destroying our planet, then we will have to be dependent on good science ... We don't even know how the simplest bacterial cell works. We want to learn what the minimum cellular components are, so we're going to be taking out all the non-essential genes. But we're also trying to design new life forms for energy production, capturing carbon dioxide or to produce chemicals.'

Venter describes his ambitious plans for creating synthetic organisms
(from an interview with Spiegel Online, 29 July 2010)

Venter began his formal education after being drafted into the US Navy during the Vietnam War. He studied biochemistry at the University of California, San Diego, where he earned his first degree in 1972, and then remained there to do a PhD in physiology and pharmacology. After that he was appointed as associate professor and later full professor at the State University of New York at Buffalo.

In 1984, Venter moved to the National Institutes of Health where he developed techniques for rapid gene discovery, and in 1992 he founded the Institute for Genomic Research, now part of the J. Craig Venter Institute with labs in Maryland and California. The institute focuses on a range of genomic research including medical applications and environmental analysis as well as ethics and law.

In 1995, Venter's team decoded the genome of the first free-living organism, the bacterium *Haemophilus influenzae*, using his new whole-genome 'shotgun' technique, a method for sequencing long DNA strands. In 1998, he founded Celera Genomics to sequence the human genome, which culminated with the February 2001 publication of the human genome in the journal *Science*.

At a White House press conference, Venter jointly made the announcement that the human genome had been sequenced with Francis Collins from the National Institutes of Health. This marked the end of a long race between Venter's firm Celera and the Human Genome Project, a government-sponsored consortium of scientists from many countries. Both groups had technically mapped the genome, but Venter's team had done it faster and more cheaply. Venter left Celera in 2002 after conflict arose between him and the company's main investor.

He and his colleagues went on to publish the first complete genome of an individual human in September 2007, having mapped his own DNA sequence. They have also sequenced the genomes of the fruit fly, mouse and rat. Another ground-breaking feat credited to Venter is the creation of the first self-replicating bacterial cell constructed entirely with synthetic DNA. They synthesized the genome for the *Mycoplasma mycoides* bacterium, with 1.08 million base pairs, using four bottles of chemicals that make up DNA, and introduced this into another cell.

The synthetic genome also included 'watermarks', coded sequences that uniquely identify it. One watermark is the code for an email address that scientists can use to report that they've cracked the code. The cell successfully divided to create millions of new *M. mycoides* cells. However, Venter maintains that this does not constitute creation of life artificially, because the cell into which the synthetic genome was transplanted already naturally contained all the necessary biochemicals such as proteins and lipids.

Venter believes that having the ability to 'write the genetic code' in this way will clarify how natural cells function, and enable cells and organisms to be instructed to perform useful tasks such as cleaning water or producing biofuels. He is also leading efforts to measure the genetic diversity of life in the Earth's oceans. His Global Ocean Sampling Expedition is assessing the genetic diversity of marine microbes in a project that has used his personal yacht, Sorcerer II. The boat sampled waters from around the world.

Venter was listed as one of the 100 most influential people in the world in *Time* magazine in 2007. He published his autobiography, *A Life Decoded*, in the same year.

MICHEL BRUNET
6 April 1940 –

---

*French palaeontologist who discovered early
hominid remains*

'You can see here that the brain case is very small, about the
same size as a chimp ... There are, of course, a lot of questions
with this new guy ... The last divergence between chimp and
human is probably older than we were thinking before.'

Brunet describes his discovery of the oldest hominid remains
(BBC interview, 10 July 2002)

Born in Vienne in Poitou, west-central France, Brunet studied
natural sciences and palaeontology at the Sorbonne in Paris, where he
earned his PhD in 1966. He became a researcher at the University of
Poitiers, where he was appointed professor of palaeontology in 1989. He
is currently professor at the Collège de France in Paris.

Much of Brunet's work has focused on searching for early hominid
fossils. In 1995, he described a new hominid (*Australopithecus
bahrelghazali*) from Chad that was 3.5 million years old. Then in 2002, he
reported the earliest human ancestor ever found. Seven million years old,
this hominid was also found in Chad and has a nearly complete cranium,
lower jaws and isolated teeth. Nicknamed 'Toumaï' (meaning 'hope of
life' in the local Goran language), the fossil was classified by Burnet as
Sahelanthropus tchadensis.

Until 1995, pre-humans had only been traced in southern and eastern
Africa. The discovery of *S. tchadensis* in north-central Africa points to a
pan-African distribution of hominids dating back at least 6 million years,
and suggests that chimpanzees diverged from humans surprisingly early,
at least 7 million years ago. However, this interpretation is controversial;
some experts have suggested *S. tchadensis* may instead be a common
ancestor of humans and chimpanzees.

GRIGORI YAKOVLEVICH PERELMAN
13 JUNE 1966 –

*Reclusive Russian mathematician who solved
the Poincaré conjecture*

'It was completely irrelevant for me ... Everybody understood that if
the proof is correct then no other recognition is needed.'

Perelman explains why he turned down the world's most prestigious maths prize
(*The New Yorker*, 28 August 2006)

Grigori Perelman was born in Leningrad (now
St Petersburg) in the former Soviet Union. He attended a school that
specialized in maths and physics, and earned a maths PhD at Leningrad
State University in the late 1980s. After holding several research positions
in Russia and the US, he returned to Russia in 1995 and worked at the
Steklov Institute in St Petersburg. A reclusive man, he avoids contact with
the press and some reports suggest he has now given up maths entirely.

Perelman published a proof of the **Poincaré** conjecture, a topology
problem formulated in 1904. In topology, mathematicians focus on the
'connectedness' of a space. They equate the shapes of a bagel and a coffee
cup, for instance, because each has a single hole and if malleable, the coffee
cup could be manipulated into a bagel shape without being cut or torn.
Poincaré called this abstract space a 'manifold' and noted that the simplest
two-dimensional manifold is the surface of a football, which has no holes
and to a topologist is equivalent to a spherical surface even if compressed.

Poincaré suggested the same is true in three dimensions – a finite 3D
space with no holes must be a sphere, in topology speak, and Perelman
finally proved this true in 2002. He was awarded the prestigious Fields
Medal and the US$1,000,000 Millennium Prize, but turned both down,
seemingly because he didn't seek fame and thought that another US
mathematician deserved equal credit.

INDEX

Quercus Editions Ltd
55 Baker Street
7th floor, south block
London
W1U 8EW

First published in 2012

A catalogue record of this book is available from the British Library

UK and associated territories: ISBN 978 1 78087 325 1
Canada: ISBN 978 1 84866 201 8

All photos courtesy of TopFoto, except p362, courtesty of Wikimedia
Editorial and layout by Hart McLeod, Cambridge
Printed and bound in China

10 9 8 7 6 5 4 3 2 1